LONGEVITY AND QUALITY OF LIFE

Opportunities and Challenges

LONGEVITY AND QUALITY OF LIFE

Opportunities and Challenges

Edited by

Robert N. Butler, M.D.

and

Claude Jasmin, M.D.

Kluwer Academic / Plenum Publishers
New York, Boston, Dordrecht, London, Moscow

International Longevity Center

ILC

Proceedings of the Congress Worldwide Revolution in Longevity and Quality of Life, held May 18–20, 1999, in Paris, France

ISBN 0-306-46315-6

233 Spring Street, New York, New York 10013

http://www.wkap.nl

10 9 8 7 6 5 4 3 2 1

A C.I.P. record for this book is available from the Library of Congress

Printed in the United States of America

PREFACE

Robert N. Butler

President and CEO
International Longevity Center

Claude Jasmin

President
International Council for Global Health Progress

Nations around the world are experiencing a spectacular increase in longevity. We are likely to live longer than our parents and to see our children grow up and grow older. Five generation families will soon become the norm. By the year 2050, 20-25 percent of the population in Western Europe, North America and Japan, and more than 15 percent of the population in developing countries will be over the age of 65. Already, 60 percent of persons over sixty live in the developing world and in the 21st century the number will rise to 80 percent. China alone will soon have 20 percent of all the world's older persons.

Society as a whole is being challenged by issues arising from this revolution in longevity, and it has become fashionable to engage in apocalyptic thinking, which is neither productive nor borne out by reality. Although the specter of the loneliness and existential suffering of older citizens is such that some people under the age of 65 find it difficult to conceive of a long-term future, persons over 85 have proven that aging does not necessarily preclude a healthy and productive life. We believe that the extraordinary progress in both curative and preventive medicine justifies optimism about the quality of life and state of well-being that can be enjoyed even in great old age. Rather than engaging in gloom-and-doom prognostication, we should look to professionals in diverse fields–science, ethics, care giving, economics, medicine, architecture, urban planning, to name a few–to develop creative solutions to the inevitable issues that will arise.

Nations must think globally, and search for ways to better serve society as a whole. While individual nations have differing cultural mindsets regarding the aged, it is clear that we need to continue to engage in dialogue on an international level. We must study current health care policies and attitudes toward the aging process, as well as movements to empower older persons in all societies. We have much to learn from each other.

Governments must prepare for the future health of their citizens by making long-term investments to educate all sectors of society in the value of good nutrition, exercise, and lifestyles that enhance well-being throughout life. People must be encouraged to take responsibility for their own health from an early age.

Obstacles will inevitably arise in the application of such a policy. Will young people adopt practices designed to lengthen their lives? Will governments be prepared to make long-term investments in costly policies for a period of ten, twenty or even thirty years, which is the time required for tangible benefits to be seen?

Of course, schools, the media and the health professions all have major contributions to make in ensuring that people, regardless of age, nationality and socioeconomic status, accept normal aging as a natural physiological process. Instead of viewing aging *de facto* as a disease process, younger generations need to learn that physical and mental decline is not inevitable. Society will have overcome a major hurdle in fighting ageism when people learn to differentiate between age-related diseases that may cause irreversible mental and physical handicaps, and the normal process of growing older.

Some economists believe that, given the need to cut spending on health, older persons need to be sacrificed in favor of the young. Yet, studies have shown that the main cause of health expenditure is serious, long-term, irreversible illness which unfortunately occurs in persons of all ages. Health care for the very old does not cost more per capita, and can often cost less than health care for younger persons. Furthermore, studies have clearly shown that older persons respond positively to medical management. A variety of treatments can reduce mortality due to cardiovascular disease, and early detection of a number of types of cancer can also save lives in this population. Unfortunately, although persons over 65 risk developing fatal cancers at a rate fifteen times greater than that of the general population, they are not encouraged to avail themselves of technological advances.

The ultimate decision will be political. At a time when the large economic power blocs are contending with tough issues of international competition, and in the context of modern economies which, typically, are run on the short term, will our leaders have the vision to commit to the long-term investments necessary for a viable health policy?

Two major socioeconomic phenomena may have a regulating effect on such economy-driven considerations. The first is the emergence of pressure groups that have come into being in response to a particular health issue–the "politics of anguish." For example, the AIDS epidemic has opened up a new domain of popular solidarity, and the disease has become symbolic of social injustice. Universal health care is a unifying issue in the 20th century and could well lead to a new form of global solidarity in the years to come.

The second major development is the emergence of ethics committees in developed nations that deal solely with health issues. We believe these committees should extend their purview to monitor the substantial growth of biotechnological science and its ability to transform human health and modern society. Society must be safeguarded against the potential for abuse of this technology, not the least of which is the potential for the manufacture of biological weapons.

At the dawn of the 21st century, the mission of the International Council for Global Health Progress and the International Longevity Center led to its organization of an international congress that brought together a variety of professionals in the field of health, to meet with associations of health-care recipients, economists, and politicians with experience as decision-makers in the field.

The common denominator was our acknowledgement that the right of older persons to enjoy good health must be a priority.

ACKNOWLEDGMENTS

The Third International Conference organized by the International Council for Global Health Progress, Paris in collaboration with the International Longevity Center in New York, was held under the auspices of Mr. Jacques Chirac, President of France, Mr. Jacques Santer, President of the European Commission, Mr. Federico Mayor, Director-General of UNESCO, Mrs. Martine Auby, French Minister of Employment and Solidarity and Mr. Claude Allegre, French Minister of National Education, Research and Technology. The volume's editors, would like to acknowledge the patronage of Espace Ethique, Assitance Publique-Hôpitaux de Paris, INSERM (National Institute of Health and Medical Research), SNIP (French Pharmaceutical Manufacturers Association). The congress was substantially supported by contributions from the following organizations: European Commission, Novartis Foundation for Gerontological Research, AG2R Prévoyance, Ernst & Young, Ipsen Foundation, L'Oreal, Nestlé France, Novartis, Pierre Fabre Santé, Rhône-Poulenc Rorer Foundation, Schering-Plough, Searle, Synthelabo, and the French Pharmaceutical Manufacturers Association (SNIP). Moreover, we acknowledge the support of the Ministry for Culture and Communication, the Ministry of Foreign Affairs, the French Embassy in the United States, Amgen, Lilly Institute, Servier and Theramex Institute. We also thank the Scientific Council and Honorary Committee for their support in the organization of the conference. The editors would like to thank Judith Estrine and Régine Boutrais for their help in the preparation of this volume and acknowledge the assistance of Milagros Marrero and Virginie Tournoud.

Finally, the editors of the English edition are particularly grateful to Searle and Novartis for their generous support.

Robert N. Butler
Claude Jasmin

CONTENTS

Introductions: Challenges and Opportunities

Global Aspects of Longevity

1. A World Wide Approach

2. Prevention of Aging

3. Industrial Responses to Aging

Quality and Quantity of Life

1. Longevity and Quality of Life

2. Ethics of Health Care for the Elderly

Social and Cultural Challenges of Longevity

1. Longevity: Generational Conflict or Interaction?

2. Aging and Cultural Diversity

3. Women and Longevity

Political Responsibilities in the Longevity Revolution

Looking Through the Mirror of Old Age

Conclusion

LONGEVITY AND QUALITY OF LIFE

Opportunities and Challenges

Introductions: Challenges and Opportunities

A MALTHUSIAN REVOLUTION

Claude Jasmin

President
International Council for Global Health Progress
Hopital Paul Brousse
Villejuif Cedex 94804
France

The International Council for Global Health Progress and the International Longevity Center, together with UNESCO, have organized this international forum to study the Malthusian revolution which has occurred in the 20th century. I have adopted this term to refer to the fact that a sixty-year-old person today can have a life expectancy of another twenty years, which could even extend to thirty or forty years in the course of the next century. Such extraordinary longevity brings with it a need for long-term planning for the older generation that is built on a sound social and economic structure. The unprecedented demographic situation brings up social, political, economic and ethical questions which we have included for discussion at this conference.

At the present time, we have managed to maintain a good quality of life with added years. The first stage in adapting society to the 21st century is therefore psychological. It is essential for everyone, no matter what their age, country or socioeconomic standing, to be taught that aging is a physiological process involving functional change; physical restrictions occur gradually and unequally and are usually first felt around the age of thirty.

The performance of the many over-85-year-olds we see today demonstrates that aging is generally compatible with a high degree of physical and social activity.

Normal aging must not be confused with disease which is, by definition, an abnormal process. Statistics in France and other economically developed countries indicate that the extension of life expectancy at birth, occuring over the past decade, has not led to an increase in the number of years of incapacity. And yet,

Longevity and Quality of Life, Edited by Butler and Jasmin
Kluwer Academic / Plenum Publishers, New York, 2000.

people under the age of 65 often find it difficult to conceive of any positive prospects in a long-term future; a quarter of the adults questioned in the United States ranked "boredom" as their greatest fear. Clearly old age still conjures up images of loneliness and depression.

Age-related diseases that cause irreversible physical and mental disabilities, such as dementia and serious functional damage, affecting, for example, sensory and motor skills, have contributed to the negative stereotype of this period of life. Sixty-five percent of a sample surveyed said they were afraid of ending up in an institution and the greatest threat for 56 percent of Americans was Alzheimer's disease. Yet the spectacular progress made in curative and preventive medicine obviously justifies an optimistic outlook regarding the quality of life and physical capabilities possible after the age of 80.

For example, current technological progress in the fields of biology, computer technology and robotics will help improve the performance and quality of life of people over the age of 70. Modern means of communication must be used to save them from isolation and loneliness. The 21st century brings with it great hopes for an enhanced quality of life for the aged.

Today we find ourselves in an ambivalent situation. We feel the need to live longer while also being afraid of a difficult old age. To cope with this contradiction, health promotion based on control of lifestyle, in particular diet and the environment, must be initiated early so as to ensure that as many people as possible can enjoy a slow and harmonious aging process. This approach must be introduced by political leaders, caregivers and health education authorities without delay so that governments can plan ahead for the radical demographic changes that will impact on the first part of the 21st century.

A pro-active, information-based approach is needed. It must be put into practice through educational initiatives and specialized social and medical facilities. The promotion of health and the prevention of disease depend on the attitude and behavior of the younger generations, on the development of medical and social structures and on older persons themselves. The International Council for Global Health Progress will play an active role in this endeavor.

To foresee the obstacles that may arise when such a policy is implemented we need to observe how difficult it is to convince the younger generation of the need for a healthy lifestyle designed to extend health expectancy and quality of life. The goal of keeping death at bay is tempered by the fear of becoming dependent, frail and lonely, and by the fear of age-related diseases, in particular Alzheimer's disease. Yet only 5 percent of the aged live in nursing homes and fewer than 10 percent of people between the age of 65 and 100 suffer from Alzheimer's disease. When one is young, it is easy to forget that 40 percent of those over 80 have good levels of physical fitness. Everyone must learn that it is never too late to begin to live a healthy lifestyle. In the United States, 60 percent of the adults surveyed said they hoped to have a long life, even to the age of 100. Americans are increasingly health-conscious and have changed their lifestyle accordingly.

But more than just technical progress is involved. A new form of solidarity has appeared among the aging population and is also supported in part by the younger generation. Concerted efforts are needed to deal with disabilities. In response to this need, associations of patients and their families, have organized around age-related diseases, such as Alzheimer's or Parkinson's. They are calling for solidarity in the form of financial support for medical and social research. The fear of being struck down by these diseases, together with the suffering and the burden patients place on their relations, provide great incentive for these movements. Infirmity can thus become a force for influencing politicians and health policymakers.

Organizing the aging population into powerful associations, with fast increasing membership, will be key to opposing discriminatory measures which may be proposed.

As I see it, the longevity revolution coming with the 21st century and the unprecedented increase in life expectancy at birth must be met with reforms of social, political and economic structures so they take into account the medical, economic, social and political consequences. These issues must be addressed from the very beginning of the 21st century, certainly in developed countries.

The first stage in the adaptation of society to the challenge of longevity in the 21st century is psycho-sociological; I see the second as being socioeconomic. The question raised by economists is whether we will have the financial resources to cope. Some economists would have us believe that the size of the older generation and the prevalence of illness amongst the aged population will have a major impact, cutting into other health budgets, and that there is therefore a short-term risk of conflict between generations over available funds. The results of the Harris poll presented at this conference run counter to this forecast and actually highlight a remarkable degree of solidarity between generations, at least in the United States. Similar observations have already been made in France.

Numerous studies have shown that a medical approach to prevention and treatment of serious or incapacitating diseases suffered by the aged can be very effective. We can only wonder then why it is so rare for this approach to be used systematically. For example, a number of treatments have been shown to almost halve the incidence of cardiovascular mortality in the aged, yet these treatments are infrequently utilized. For cancer, where early screening is often proven to be effective, we can see that the risk of developing cancer is eleven times greater in the over-65 age group and the risk of death is fifteen times higher than in younger age groups. Yet the aged benefit very little from such medical progress, even in economically developed countries.

It will take between 10-30 years to reap the benefits of a government-sponsored disease prevention program. Will governments make a commitment to long-term investments as part of a costly prevention policy over a ten to thirty year time-span? What will we do in the year 2030 when one person in five in these economically developed countries is over the age of 65? What recommendations will be made by economists concerned with the prospect of soaring health costs?

The real decisions will be political. How will health decision-makers agree on the social and economic choices that need to be made to meet the challenge of human longevity in the 21st century? At a time of tough international competition between the leading economic blocks, will decisions be made on the long-term investments that are needed as part of a comprehensive health policy, rather than choosing short-term options which are a feature of modern economies? What will be the attitude of our political leaders to work-sharing at a time when older persons are perfectly capable of efficient work for another ten to fifteen years after the present retirement age? Are these social and economic aspects compatible right now?

Two major sociological phenomena may have a regulatory role to play in the face of such economic pressure. The first, and as I see it, the most important, is the emergence of associations of older people suffering from, or directly affected by a specific medical condition. More of these associations are being set up as pressure groups, capable of lobbying both the general public and politicians.

The AIDS epidemic has given rise to a new area of public solidarity. The disease has become the physical manifestation of an injustice, and is symbolic of all forms of social injustice. The right to health, which is a rallying cry in the latter part of the 20th century, could thus generate worldwide solidarity between the aged and, in an ideal situation, include people in developing countries where more than half of the world's population over the age of sixty lives. Such "health advocates" could also find backing in environmental movements.

In the most favored nations, we have seen the emergence of ethics committees dealing solely with health issues and this stands as the second major development in the late 20th century. These ethics committees play a role as arbiter in any conflicts of interest arising between patients and their family or patients and public authorities. Their scope and prerogatives must be extended because of the huge development internationally of biotechnologies capable of changing the health of human beings.

The International Council for Global Health Progress plans to organize an international conference at the turn of the century to bring together various health professionals and health users, in particular associations of patients, economists, politicians and figures of moral authority. The rights of the aged to health will be a key part of the study.

I would like to conclude by saying that the Malthusian revolution has brought with it the seeds of a new ethical code of respect for human vulnerability and can thus play an important role in building a new era of human wisdom.

A GLOBAL APPROACH

Maurizio Iaccarino

UNESCO
7 place de Fontenoy
Paris 75732
France

The phenomenon of aging is of concern to every individual and every society. We do not want to die because we do not want to abandon our relatives, our friends and, especially our plans for the future. The latter phenomenon is very well known to psychiatrists who study the aged from the point of view of their capacity to have future expectations.

We do not want to die. When mammalian cells were first grown in a laboratory more than 60 years ago, Alexis Carrel concluded that chicken fibroblasts cultured *in vitro* were immortal. Although this interpretation was due to an artifact, it raised at that time an enormous issue: can we avoid aging? This question has always been with us.

Governments throughout the world have supported medical care and medical research in order to improve our life and protect us from disease. These efforts have been very effective.

We now live better and longer lives, although there are striking inequalities between developed and developing countries. But, this change poses new challenges to the scientific community and to governments. It is increasingly apparent that there is a pressing need for an integrated study of the phenomenon of aging, not only because of its far-reaching implications for health care provision, but also, and predominantly, because of the major social consequences.

UNESCO has been entrusted by the world community with the promotion of international cooperation in science, education and culture. The most important aspect of our programs is their interdisciplinary orientation. We work for our member nations and report the results of our activities to them. It is therefore particularly appropriate that your Conference be held here at UNESCO House, since it is to the attending decision-makers that you wish to bring the results of your work.

In June 1996, an international conference entitled "Human aging: adding life to years" was organized by UNESCO. It focused on the biological basis of aging and on the possibilities of improving the length and the quality of life. Your own conference will not only deal with biology, pharmacology and neurobiology, but also with demography, as well as social and cultural problems. Your forum has

Longevity and Quality of Life, Edited by Butler and Jasmin
Kluwer Academic / Plenum Publishers, New York, 2000.

brought together high-level experts from a wide range of disciplines, and presents an opportunity to discuss the phenomenon of aging from a multitude of angles.

The recent advances in the study of the basic processes of senescence have produced many changes in the field of gerontology, which now goes beyond the earlier preoccupation with age-related diseases, such as cancer, Alzheimer's disease and dementia. The new gerontology deals with the progressive non-pathological changes that occur with aging and that influence the biological and physiological status of human beings. The study of telomeres, free radicals, genes controlling the cell cycle and so on are expected to yield results that will throw light on the basic biological mechanisms that determine longevity.

In addition to fundamental research, a second major aspect of the new gerontology recognizes that there is more to successful human aging than just the avoidance or delay of disease: it also requires the maintenance or enhancement of physical and cognitive functions. Interdisciplinary studies on the elderly, combining physiology, epidemiology and the social and behavioral sciences, have identified lifestyle, nutritional, psychosocial and other factors important in maintaining or improving physical functions. These include strength and balance, cognition and memory. It is therefore evident that a multidisciplinary approach is required to study and solve this problem. Other important factors, such as loneliness, isolation, lack of security and lack of mobility must also be taken into account. In old age, loss of muscle strength resulting in frailty is a major determinant of independence.

Living longer and "becoming old later" is a real revolution in the history of humankind. The percentage of old people in the population is increasing throughout the world and this phenomenon brings with it new problems of a socioeconomic nature. We need a global approach to the study of longevity, an approach which includes demography, epidemiology, sociology and economics. We need to involve all the professionals in these fields in discussions with the decision-makers in our societies if we are to meet the challenges ahead.

I am sure this Congress will match the expectations of its organizers and will make real progress in the search for an appropriate and equitable treatment of the aging population in our society.

LONGEVITY AND WOMEN'S HEALTH

Nafis Sadik

United Nations Population Fund (UNFPA)
220 East 42nd Street
New York, NY 10017
USA

INTRODUCTION

As the organizers of this conference have correctly recognized, we are indeed in the midst of a "worldwide revolution in longevity" as a result of demographic and health trends in every region. Many aspects and implications of this revolution have not yet received the attention they deserve from policy makers, including issues related to the health and human rights of older women.

First, let us consider the magnitude of the issue. Older populations are growing in absolute numbers and proportionately in both developed and developing countries. Today, according to the UN Population Division, there are more than 578 million people over 60, and this generation is growing at an unprecedented rate.

In the more developed regions, the proportion of the population above 65 has increased from 7.9 percent in 1950 to 13.5 percent today and is projected to reach 24.7 percent by 2050. In parts of Europe, North America and East Asia where low fertility and long life expectancy are the norm, the proportion of people above age 65 is rising more rapidly than any other age group. The proportion will double within the next 35 years—approaching or exceeding 40 percent in Japan, Germany and Italy. In some countries, populations above age 85 will more than double in the same period of time.

Many developing countries will have to contend with both a rapid increase in numbers of older people and continued high population growth. The number of older persons being added to the world's population is now approaching 9 million per year, about 10 percent of the total annual increase. This will increase to over 14 million a year by 2010 and 21 million a year by 2045. Currently developing regions account for 77 percent of the increase in the older population; this share will grow to 80 percent by 2015 and over 97 percent by 2050 (when more than one quarter of the growth will be in India).

Africa and Western Asia still have relatively low proportions of older people as a result of continued high fertility. In sub-Saharan Africa the proportion

Longevity and Quality of Life, Edited by Butler and Jasmin
Kluwer Academic / Plenum Publishers, New York, 2000.

above 65 will increase only from 3.2 percent to 4.2 percent between 1995 and 2025. However, the number of people over 65 will increase rapidly, from 22.7 million to 61.2 million.

There is also considerable variation within regions, reflecting the different pace of fertility and mortality in different countries. Many Caribbean countries, for example, have high proportions of elderly populations, quite unlike their close neighbors in Central America.

SUPPORT SYSTEMS

As numbers of elderly people increase, traditional family support systems in much of the world are eroding as a consequence of urbanization and other modern pressures. This situation calls for new arrangements for elder care. Even developing countries where the proportion of older people is still small need to plan ahead. Most of the burden of caring for older family members has fallen on women; as the burden grows heavier, national policies must respond.

Formal retirement age in developing countries is usually lower than in developed countries, but relatively few older people occupy formal sector jobs in which the concept is meaningful. More countries need to develop public and private pension plans with wider coverage and workable social security schemes, learning from the experiences of the industrialized countries and some developing countries. Families also need to be assisted in caring for elderly members.

HEALTH CARE

Better health care is also essential, to enable older people to remain healthy, independent and productive longer. Health policies need to concentrate on prevention rather than medication and surgery to address conditions related to aging, and to pay more attention to health promotion, healthy lifestyles, environmental health and occupational hazards.

Let me turn to the issue of older women's health. A majority of older persons are female, and yet their health concerns have not received the attention they deserve from policy makers. This reflects the larger problem of inadequate attention to women's health needs throughout the life cycle.

Women generally live longer than men, but they can also expect more years of ill health late in life. Many older women suffer from chronic health problems caused by years of neglect, discrimination and the hardships of their childbearing years.

Biological factors that lead to disease and disability in women can be exacerbated by lifelong discrimination. In many poor households and communities, women work harder than men but eat less. Women and girls have unequal and inadequate access to basic services and education. They may be exposed more than men to environmental health risks, as a result, for instance, of their more frequent contact with polluted water, indoor air pollution from smoky kitchens, and lifelong exposure to pesticides, all of which threaten women's health.

Gender discrimination is also closely related to women's greater likelihood of being poor, which in turn contributes to women's ill-health at all ages.

For instance, poor nutrition, anemia and certain vitamin and mineral deficiencies are linked to heightened risks of maternal mortality and morbidity, obstructed or prolonged labor, stillbirths and miscarriages.

Many long-term female health problems originate during the reproductive years. Their extent and severity are closely tied in one way or another to the denial of women's reproductive rights. These encompass both the right to receive adequate health care, including the means and information to control one's fertility, and the right of free choice in matters of sexuality and childbearing.

Each year, nearly 600,000 women in developing countries die from causes related to pregnancy and childbearing; a third of these deaths can be attributed to unwanted pregnancy and the failure or inadequate availability of contraceptive services. Many more women survive but suffer serious, lifelong health consequences. Better maternal care and wider access to reproductive health services, including family planning, could prevent many of these deaths and greatly reduce maternal morbidity.

While the average age of marriage has increased in much of the world, in a number of countries social pressures still lead girls to marry and begin childbearing soon after puberty. This puts them at greater risk of death and illness during childbearing than the risks faced by older women, and has implications for health in later life. In most societies, young people are increasingly at risk of sexually transmitted disease, including HIV/AIDS, and behavior that leads to poor health later in life, including smoking and substance abuse, which often begin in adolescence.

Multiple and closely-spaced births can result in diabetes, and continuing pregnancies into later life have been shown by many studies to have damaging effects on women's health.

Sexually transmitted diseases often have long-term effects. For instance, the human papilloma virus has been implicated as a cause of cervical cancer. STDs could be prevented much more effectively, and their impact greatly reduced if reproductive health information and services were more widely available, and if women were able to exercise their right to have control over their sexuality free of coercion, discrimination and violence.

A life-cycle approach to women's health, combined with actions to redress gender inequality, will provide the opportunity to extend the years in which women enjoy good health.

Healthy adolescent development, including the postponement of marriage and childbearing and avoidance of STDs, has lasting health benefits. Appropriate reproductive health information and services need to be made available to young people, married and unmarried.

Integrating comprehensive reproductive health services into health-care systems improves the lifelong health prospects of both women in their childbearing years, and those of their children. The 1994 International Conference on Population and Development articulated the need for universal access to a package of reproductive health services, including family planning, pre- and postnatal care and safe delivery, prevention and treatment of infertility, prevention of the need for abortion and management of the consequences of abortion, treatment of reproductive tract infections, prevention of sexually transmitted diseases and other reproductive health conditions, and active discouragement of harmful practices such as female genital mutilation.

A life-cycle approach to women's health would also increase the research and policy attention paid to issues related to menopause. Physical effects of changing hormonal levels at menopause include increased risk of heart disease, increased risk of reproductive tract infections, osteoporosis, and an acceleration of aging in skin, muscles and other tissues. Women with osteoporosis are more likely to fracture bones; 80 percent of hip fractures at older ages are in women. Psychological changes include a loss of sexual appetite and pleasure, mood swings and depression, which may also be influenced by changing social identities and self-perception.

In many countries menopause is simply not given priority as a reproductive health issue by health service providers, although this is changing. Some developing countries, such as Chile, have already initiated programs to incorporate menopausal information and services into their health systems.

For women and men, longer life does not necessarily mean more years of inactive, impaired or disabled life. As life expectancy has increased, so have the years of healthy and active life. In developed countries (where the data are better) these increases have been dramatic even in short periods of time. Chronic conditions like arthritis increase with age, but they need not impair normal activities or lead to disability.

While disability becomes an increasing concern at the oldest ages, it should not overwhelm other health policy concerns, including prevention or early correction of conditions which contribute to later disability.

Until recently, health care for the aged erroneously assumed that the health concerns of women converged with those of men after menopause. This is starting to change. As a matter of fairness, greater service priority should be given to addressing osteoporosis and chronic degenerative diseases in older women. Another issue that should be addressed in many countries is the limited contact that many women beyond reproductive age have with health services, delaying detection of conditions like breast and cervical cancer.

Differentiation of health statistics according to gender, and monitoring of gender differences in susceptibility to various conditions are needed to improve preventive efforts.

Similarly, greater priority should be given in medical research to interventions that could increase the healthy life expectancy of women; these include the use of drugs to address genetic factors, better health through sensible caloric restriction, and reduction of obesity and other diets that result in malnutrition.

There are many other important research and service issues affecting older women's health:

- Little is known about gender-specific approaches to mental illness among older women, especially depression and dementia.
- Prevention of chronic disabling conditions, including arthritis, osteoporosis and incontinence, should be directed at younger age groups as we learn more about causes and risk factors.
- Data are needed on how lifetime morbidity and low social status impact on the health status of older women, on the health of women working in dangerous conditions, especially in the informal economy, on the domestic workload, and on domestic violence.

HUMAN RIGHTS

Finally, far greater attention must be paid to the human rights of older women—including redressing the inequality they face with respect to pensions and inheritance, and preventing violence against them.

CONCLUSION

It is imperative that we address emerging issues for older persons in the context of policy decisions. It is imperative that developing countries do their part to ensure that policies for the elderly are initiated and that the health and social well-being of older persons take their rightful place in the provision of social services. These efforts are essential if we are to create a future in which, as people live longer, they can enjoy their added years and make the maximum possible contribution to their families and society. I am confident that this meeting will play an important role in alerting policy makers to this important concern and its ramifications for human development.

OPENING REMARKS

Bernard Kouchner

Minister of Health
4 avenue de Segur
Paris 75350
France

The keys to one of the major challenges of the next century were given to us by Cicero, in his *Cato Maior de Senectute*:

"One must fight old age and combat its inconveniences with the weapon of care; one must resist it as one resists disease; maintain one's health, exercise as appropriate, eat and drink to regain one's strength without undermining it. But it is not enough to take care of one's body; one must also nourish the spirit and the soul. Indeed, both run the risk of being snuffed out by old age like the flame of a lamp deprived of oil."

By the year 2020, the world will be home to more than one billion people over the age of 60, 710 million of them in developing countries. In a recent report, the World Health Organization pointed to the worldwide increase in longevity, even in the poorest countries such as Sierra Leone, where a 13-year increase in life expectancy is predicted by 2025. In 2025 in China, people over 60 will outnumber the entire population of the United States today.

In turn-of-the-century France, only one person in four reached the age of 65. After the Second World War, this figure rose to two out of three. Today it stands at four out of five. "How dramatic! What a burden!" are the reactions we sometimes hear.

"What an achievement, what a victory over illness and death," we should answer back. That is, as long as we are able to address the key question: how can we make these extra years better years? How to reconcile improvements in life expectancy with quality of life? And how to reduce the inequalities we see emerging in terms of increased life expectancy?

Here too, the same old story of unequal progress is repeating itself. And once again, the most vulnerable are those who gain the least. The rich get richer, as the saying goes. Social class and gender divide us once again. Women benefit more than men from increased life expectancy, but also face an increased incidence of

Longevity and Quality of Life, Edited by Butler and Jasmin
Kluwer Academic / Plenum Publishers, New York, 2000.

disability. As for those least fortunate economically, they face both slower progress on the life expectancy front and higher disability rates.

The unprecedented evolution of the age pyramid is indisputably one of the defining features of the 20th century, along with the generalized spread of industrialization and urbanization. This change has occurred without any real warning, none of the major social prophecies of modern times having perceived, I believe, the true extent of its impact. And it is probably irreversible, at least in wealthy countries. Uncertainty remains in developing countries, where they are often helpless to stop the terrible advance of the AIDS epidemic, which has already begun to cancel out and even reverse this trend in nations most severely affected.

For their part, however, Western and Japanese societies seem destined to become societies of the elderly due to the dual impact of increasing life expectancy and falling birth rates. Should we be pleased? Need we be alarmed? Above all, we need to be more aware and to take into account the cultural shift taking place as new generations of retirees aspire to live their final years to the fullest extent. This includes a satisfying sex life, as the recent fantasy-driven excitement over the now-famous Viagra pill has shown.

This enthusiasm raises with new urgency and scope the question of how far society is willing to go to bring the old Faustian myth to life. As Francois Forette, one of the top French specialists in the field, points out, "there are more and more elderly people, but they are aging less, and aging in good health." Indeed, we may now hope to grow old in good health by preventing pathological aging.

You will no doubt express your ideas more clearly than I during this convention, but I would like to raise with you four vitally important avenues for exploration.

The first concerns the slowing of biological aging. This objective, although promising, remains primarily a research issue, and I will not explore it any further here.

The second avenue is that of reducing risk factors that lead to chronic or debilitating illness. We know of the damage caused to young adults by smoking, lack of exercise, obesity, etc. Their consequences are delayed, normally striking after the age of 60. The way we live determines, in part, how we age. Of course, the aging process begins long before we become "senior citizens." Aging is not just for old people. Aging affects everyone from the day they are born. It is a process concomitant with growing up, in short, with life itself. It means that preventive action must be taken early on, which brings us to the issue of health education and prevention, the poor relations of public health. We must give them the attention they deserve, for they have a major role to play.

The third avenue is the early detection of treatable asymptomatic diseases. Indeed, we now can clearly distinguish between pathological aging–with its cortege of diseases that we are better and better equipped to prevent and treat–and inevitable physiological aging.

We now know that it is not age itself that diminishes our main vital functions, but rather disease. Although disease strikes more frequently with age, preventive options already exist in many cases:

- First, cardiovascular and cerebrovascular illness, the world's leading cause of death and disability for people over 65, can be prevented by treating problems like hypertension and cardiac irregularities.

- But there is also hope for osteoporosis, malnutrition, domestic accidents, etc.

Osteoporosis is now a major public health issue in terms of the growing number of women affected as well as the seriousness of its consequences and its repercussions on patient quality of life.

This situation led France to take a close look at the problem. It resulted in a ten-year government plan being released in Fall, 1998, with the goal of reducing bone fractures among the elderly by 25 percent within 10 years, and by 50 percent within 25 years. The program is being developed in line with recommendations made in the report called *Osteoporosis: Action for Prevention* which was submitted by a committee of European experts chaired by Professor Blanchard.

The fourth avenue consists of "delaying" the loss of functional autonomy by the elderly. Major progress has been made in understanding the mechanisms that lead to the loss of autonomy.

It is important to identify three areas: biomedical change (sensory, cognitive, osteoarticular, sphincteral and other deficiencies), their functional consequences (inability to accomplish daily tasks) and social disadvantages (need). Each of these areas can be assessed individually. It is vital to fight the idea of a total dichotomy between two worlds and two radically different and separate life stages: idealized autonomy followed by dependency as sudden as it is total.

Of course, reality is much more complex. Our lives follow courses interwoven with multiple and constantly renewed affective, psychological, physical, social and economic interdependencies. As Doctor Marie-Pierre Harvey puts it: "Throughout our lives, we are led to seek balance between our aspirations, our abilities and the means available to us and those around us. This balance is regularly thrown off kilter, and we develop strategies to find new balance suited to our new circumstances. Dependency is not an unworthy state, but something that follows us in various forms all our lives, and for which we constantly adjust."

Preventing pathological aging requires more than simply a medical approach. "Health," wrote the philosopher Canguilhem, "is the prolonged ability of an individual to physically, emotionally, mentally and socially face his environment." Health, well-being, and quality of life can only be envisaged from a holistic approach. This is a difficult and complex task, but one which has the capacity to restore to the field of medicine the humanistic element that the dazzling biotechnological developments of the past thirty years have overshadowed and shunted aside.

Lastly, policy on aging can be no more than a magnified reflection of society as a whole. A welcoming, community-oriented society that respects its members will show these same virtues in the way it treats older persons. In contrast, an individualistic society dominated by corporate interests will draw up a policy on aging in its image: cold, segmented, exclusive.

We in our societies of egotistical and uncaring abundance would do well to look for inspiration to the societies of the developing world that still know how to treat their elders with love and respect. Because they are the guardians of memory and culture, but also simply because they are human beings.

As Simone de Beauvoir's lovely aphorism puts it: "What should a society be in order that a man, in his old age, remain a man? The answer is simple: he must have always been treated as a man. "

WORLDWIDE REVOLUTION IN LONGEVITY AND QUALITY OF LIFE

Alain Pompidou

Laboratoire d'Histologie Embryologie-Cytogenetique
Medical Faculty
24 Rue de Fbg Saint Jacques
Paris 95014
France

Human longevity and, hence, the aged population is increasing rapidly. In 1900 there were more than 30 million 65 year olds globally. They will be ten times more, 350 millions, in 2000, and probably 50 times more, 1.5 billions, in 2050. In Europe, Japan and the United States, 20 percent of the population is more than 65 years old. In the European Union, by 2025 we will have 10 million young people under 20 compared to 40 million retired people over 60.

The aging of the world population, and a decline in births (probably greater than expected according to the last OECD statistics) generates tension in a society caught between the development of leisure in industrialized countries (and to some extent in new emerging countries) and the poverty that exists in whole sectors of the world. This unbalance is enhanced by an acceleration of the great migratory flux, which destabilizes people and society as a whole. So, the changes in the age structure of populations leads us to question traditional socio-cultural references. In this context, demographic research must be continued to expand our knowledge of population dynamics. Research is also needed on the biology of aging and on the social, economic, cultural and ethical aspects of the world population's aging.

The United Nations will mobilize on behalf of the elderly in 1999. Today's congress will help identify what must be undertaken. The issues include answering to the massive needs of aging populations, especially in industrialized countries; answering to the demand for a better quality of life; introducing economic debates in a context of respect for cultural and ethical values.

We must engage in a new international solidarity on behalf of the elderly, aiming to prevent the effects of aging as early as possible and to find answers to this new issue, the revolution in longevity. Information and communication technologies can enable us to maintain aged people in active lives at a relatively low economic cost. As longevity has now gained a new economic dimension, three questions

Longevity and Quality of Life, Edited by Butler and Jasmin
Kluwer Academic / Plenum Publishers, New York, 2000.

emerge: What are we going to do? What are we willing to do? What can we do? These are the questions we have to confront and discuss today.

Global Aspects of Longevity: 1. A Worldwide Approach

THE REVOLUTION IN LONGEVITY

Robert N. Butler

International Longevity Center
60 East 86th Street
New York, NY 10028
USA

INTRODUCTION

A silent and unprecedented revolution in longevity has occurred in the 20th century. The industrialized world has gained over 25 years of life, and this achievement is nearly equal to the life expectancy attained during the preceding 5000 years of human history. Nearly 20 percent of this gain is from base age 65.

Longevity is not a function of genetic evolution, but of social and economic progress, better public health and nutrition, and the application of important medical concepts, such as the germ theory of disease. Furthermore, in this century we have enjoyed steady success in meeting the challenges posed by the increasing number and proportion of older persons in many nations, through such advances as Social Security, health insurance, and biomedical research.

Thanks to the new longevity, society has experienced many positive developments. These include increased productivity as a result of the reduction of disruptive illness and premature death, and the emergence of the multigenerational family. Moreover, the new longevity has led people to plan even more for the future. New savings are being generated through pensions, which constitute one primary source of capital formation today. The numbers of older persons have been the catalyst for basic research and clinical investigations into aging. There have been notable drops in disability rates and improved quality of life.

The aging revolution is also contributing to a transformation of our health care and service delivery systems by requiring a more comprehensive, integrated approach to patient care, by the development of new technology to deal more effectively than we presently do with the frail and bedridden, and by advancing long-term care programs for all ages and conditions. The revolution in longevity is

Longevity and Quality of Life, Edited by Butler and Jasmin
Kluwer Academic / Plenum Publishers, New York, 2000.

also forcing us to rethink end-of-life decisions. Finally, it has stimulated consideration of ethical and philosophical aspects of longevity, aging, dying, and death, and the equitable allocation of resources among the generations. But the adaptive responses required by the new longevity are incomplete as we approach the 21st century.

In the U.S., for example, the "baby boomers," born between 1946 and 1964, will begin to retire in the year 2008 and receive Social Security benefits. In the decade 2020 to 2030, they will make up 20 percent of the U.S. population; in other words, one of five Americans will be over 65. Baby boomers, now comprising one-third of the nation's population—the largest generation in U.S. history—were an unexpected phenomenon. Initially, there were not enough diapers. Then, there wasn't enough space in public schools, or room in college, or jobs, or housing. Without planning now, the U.S. will also be unprepared for the baby boomers as they enter the final portion of their lives. Thus, they constitute a generation at risk. The United States is not unique. Japan will arrive at the point where 20 percent of the population is over 65 a full decade ahead of the U.S. Europe, too, is aging rapidly.

However, when we look at the developing world, we find that the longevity revolution is incomplete and this is very disturbing. Within this population, life expectancy is significantly lower than in the technologically advanced nations of the world. Ten percent of lives are lost to disability and disease. We need to address this disparity and work to develop the economic and social conditions in which an aging society can flourish. We await the democratization of longevity. Nonetheless, sixty percent of older persons are in the developing nations, and in the next century it is anticipated that the percentage will rise to 80 percent.

The longevity revolution is a derivative of the industrial-scientific revolution, and, like this revolution, fundamentally affects everything: the family, economic productivity, the health care system, culture, and so forth. All major changes have unpredictable ramifications. For instance, the industrial-scientific revolution, which drastically transformed societies in positive ways, also contributed to adverse environmental changes, such as global warming and deforestation. Similarly, along with the longevity revolution has come a significant minority of severely disabled and impaired people, devastated by Alzheimer's and vascular dementias, frailty, and problems of mobility. Both the industrial-scientific revolution, now about 200 years old, and the longevity revolution, less than 100 years old, have occurred with great rapidity. It will take time to complete adaptations to the longevity revolution which will require imagination, skill, and the fair distribution of resources.

ISSUES OF LONGEVITY

However, this demographic change has also led to considerable cultural, political and economic uncertainty as nations consider rising public pension and health care costs. Can we afford older persons? Are they a burden, given pension and health care costs? The welfare state is under attack and four concerns are widely voiced:

- How much *can* and *should* society and the individual bear?
- Will costly medical and financial dependency take away resources from the young and create intergenerational conflicts?

- Will the overall aging of the population, its burdens and costs, cause stagnation of the economy? What of society at large, in war and peace? Will it cause a weakening of the national will?
- Will excessive power concentrate in the hands of older persons? Will we live under a gerontocracy?

Underlying these four concerns is anxiety about aging based upon the human fear and dread of growing old, becoming dependent, and of dying and death. Our anxiety creates prejudice against "them" in order to distance ourselves from the realities of late life. I have named this phenomenon *ageism*, and the word is now part of the English language.

However, the future clearly belongs to those who are prepared. We must overcome our fears and ask what should be done in response to the longevity revolution. Here are a few suggestions:

1. Implement and enforce present policies, especially social protections that would help free old age from poverty and, in heterogeneous societies, overcome differences in longevity due to ethnicity, race and lower socioeconomic status. Special attention must be given to the situation of women, who live longer, carry most of the care giving burden and are at greater risk for chronic disease, poverty and institutionalization. Over the past century, we have seen social protections evolve in Europe, Japan, and the United States. Following the Second World War, especially, we saw the rise of the welfare state, now in various degrees of reorganization, reform and retrenchment. Unfortunately, the developing world has few such social protections.
2. Political, economic, societal and even personal changes are also underway in response to the new longevity. These changes require an alteration of our mindsets, and a new and constructive vision of a maturing society.
3. Print and electronic media have a great obligation—and opportunity—to educate the public and its decision-makers. We must especially appreciate the varied responsibilities of the generations. How do the young of today look upon their parents and grandparents? How sensitive to the future generations are the middle and older generations. We must create new images and roles for older persons. (Will grandchildren ever look at their grandparents the same way when the great astronaut, John Glenn, returns to Earth?)
4. The world of work and work-family policies must dramatically change. Since we are living longer, people will have to work longer and be promoted later. Job-sharing will become increasingly important, as will sabbaticals for everybody, to ensure continuing education.
5. Understand that the answer to the question, "Who is responsible for our old age?" is: the individual, the family, the community, the economic enterprise, *and* government at all levels. In other words, such a major transforming change in society requires participation of all persons and institutions. This may be particularly true of medicine, which, having contributed to the new longevity, has the responsibility of intensifying its commitment to helping society adapt.

The world is changing. As we move into the global economy, we require a healthy, vigorous, and educated population to remain both competitive and cooperative. We need to create social policies that address the new longevity, and we cannot overlook the fact that all policies, all services, all means of income maintenance ultimately depend upon, and can only be sustained by, a vigorous economy.

ADDRESSING THE ISSUES

Initiatives to address the issues posed by the longevity revolution may be divided into two broad categories: (1) socioeconomic policy, and (2) health and social service system development. A successful socioeconomic initiative depends upon fostering societal and individual productivity, including the productive utilization of older persons—what I call productive aging; the development of multi-tiered income maintenance, including a universal and intergenerational Social Security system to assist persons of all ages, as well as the opportunity for personal planned savings and investment; and construction of pro-family policies, such as daycare systems, in keeping with the new and changing family structure. Already, vast sums of private wealth pass from old to young, not from young to old. Ultimately, the critical impetus for continuing productivity is investment in scientific research and development, in education, and in improving health.

A successful health and social service system must provide universal access. This system should be comprehensive and promote collaborative care. It needs to elevate the roles for social services and the so-called allied health provider systems which are so critical to older persons. Such a system must emphasize the responsibilities of the patient's family and community, and focus on health education, health promotion, and disease prevention. It should also develop a strong biomedical and behavioral research strategy that emphasizes gerontology, and longevity science that exploits cell and molecular biology.

Gerontology comprises not only research on aging, but also research on longevity. We must capitalize upon the "new" biology—from telomeres that clock aging, to new approaches to Alzheimer's disease and the other disorders of longevity. Today, specific aspects of aging and aging-related disorders are already mutable, through biomedical and behavioral interventions. Options such as hormone replacement therapy, weight training, exercise and healthy diet, make it possible to extend the healthy and productive years of life.

We must build public health programs that encourage health habits to enhance both the quality and quantity of life—a prescription for longevity. Soon, with the completion of the *Human Genome Project*, we will be able to build a more individually customized means of helping to prevent disease.

We have seen more changes in this century than in all of human history. We have seen, among other things, the extraordinary new longevity on the one hand, and, on the other, the potential for the destruction of humankind. There are serious problems: the emergence of new diseases, such as the Acquired Immune Deficiency Syndrome; the recrudescence of past illnesses, such as poliomyelitis; the failure of public health and of occupational safety and health to adequately protect people; and environmental damage, such as the hole in the ozone layer, which may increase the incidence of skin cancers and cataracts. The longevity revolution cannot be taken for granted. We have already seen the loss of life expectancy in some countries, reminding us that longevity cannot be taken for granted, and, like liberty, must be guarded vigilantly.

Gerontologists and ambitious molecular geneticists, as well as other biologists, are considering possible germ line interventions to change the course of human destiny, even extending the life span. In 1898, H.G. Wells wrote *The War of the Worlds*, which stimulated the imagination of the young Robert Hutchins Goddard. Goddard later established the foundations for interplanetary space travel. Now 100 years later we may be poised to stretch the limits of biological time.

CONCLUSION

This brings me to my final points: To what societal and individual purpose should humankind use new gains in longevity? What are the appropriate philosophical and ethical considerations? How should we conduct our lives? In Western civilization, there has been a relatively modest theological, philosophical and literary output concerning old age and longevity, perhaps because in the past, old age was attained so infrequently. There is *Ecclesiastes*; Cicero's *De Senectute*; some of Montaigne's essays; an indication of "stages of life" in Shakespeare's plays and Rousseau's writings; the thoughts of Elie Metchnikoff, who coined the term *gerontology*; the remarkable play of George Bernard Shaw, *Back to Methuselah*, and, perhaps the greatest love story ever of old age, Gabriel Garcia Marquez' *Love in the Time of Cholera*. In the great civilizations of the East, there may be much more that has been written which is pertinent to the revolution in longevity.

I personally believe that we should bring together the finest thinkers of the East and West to address the issues created by the era of longevity. Older persons themselves should take the lead—pioneering in an exploration of this virtually new stage of life.

We cannot wait until the year 2010, when the world is significantly older, to alter our social policies, create an effective health and social services system, and further the quality of life. Indeed, from economic, societal, familial, cultural, and philosophic perspectives, we have a relatively short time to prepare.

LONGEVITY AROUND THE WORLD

Vladislav V. Bezrukov

Institute of Gerontology
Academy of Medical Sciences of the Ukraine
Vyshgorodskaya Street, 67
Kiev 254114
Ukraine

INTRODUCTION

The world population is aging. We are witnessing the global increase in average life expectancy at birth, in numbers and percent of people aged 60 years and over (higher than for the entire population) and an even faster increase in numbers of the oldest old population.

LIFE EXPECTANCY

An unprecedented gain in the average life expectancy at birth has occurred during the 20th century (double that for the previous history of mankind). This trend was also sustained during the last decades, both in developing and developed countries (4,7) (Fig.1).

At the beginning of 1950s, life expectancy of the world population was 47 years. In 1995-96 it was close to 66 years, ranging from 40-42 years in Uganda and Zimbabwe to 79-80 years in Japan, Canada, and Australia, with China, Russia and India in between (8) (Fig.2).

In most developed and developing countries life expectancy is increasing (in more developed countries by 4-5 years during the last two decades, and in developing countries up to 10-12 years). The Eastern and Central European countries are exceptional in their lack of improvement in life expectancy, and in some countries of the former Soviet Union life expectancy decreased 4-5 years since 1989.

According to estimates, the trend of life expectancy increase will continue well into the 21st century. From 1995 to the year 2025, life expectancy for developed countries will rise: in males from 71 to 76 years, in females from 78 to 82 years (7). Corresponding figures for developing countries are: for males 60 to 66 years, for females 63 to 73 years. The gain in life expectancy will be higher for the

Longevity and Quality of Life, Edited by Butler and Jasmin
Kluwer Academic / Plenum Publishers, New York, 2000.

developing countries of Asia, Africa and Latin America. For example, expectation of life at birth in Africa, which is now over 53 years, in 2025 will be around 66 years and 73 years in 2050, almost a 40 percent increase (5).

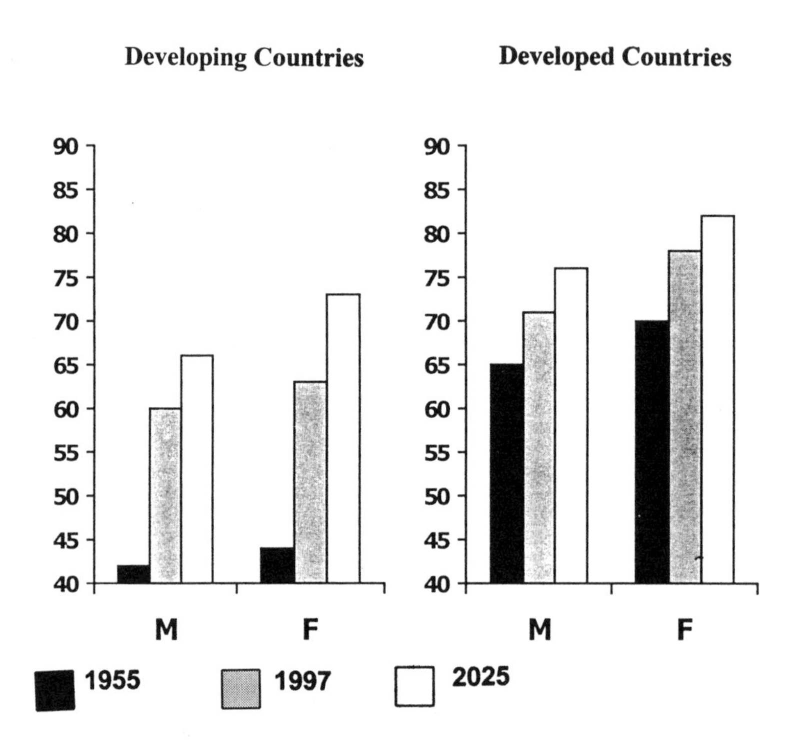

Source: United Nations, 1995 and U. S. Bureau of the Census, I. P. C., 1996

Figure 1. Years of Life Expectancy at Birth 1955, 1997, and 2025

AN AGING WORLD

The percentage of elderly people is increasing in almost all countries of the world (15,17) (Fig.3). At present, the elderly make up 15 percent or more of the entire population in 45-50 countries of the world, and this percentage is over 20 percent in 11 European countries (Belgium, Bulgaria, Denmark, France, Germany, Greece, Italy, Portugal, Spain, Sweden and UK) as well as in Japan (8).

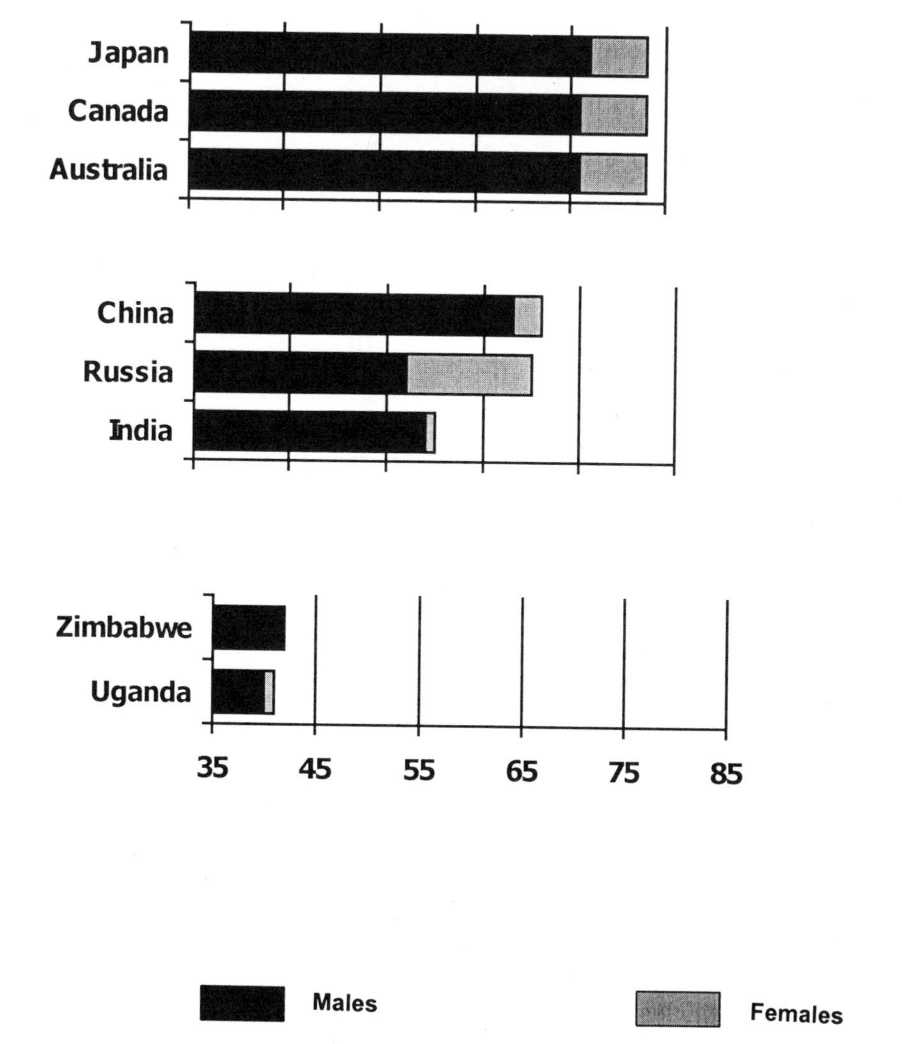

Source: United Nations, 1995 and U.S. Bureau of Census

Figure 2. Years of Life Expectancy at Birth in Various Countries, 1996

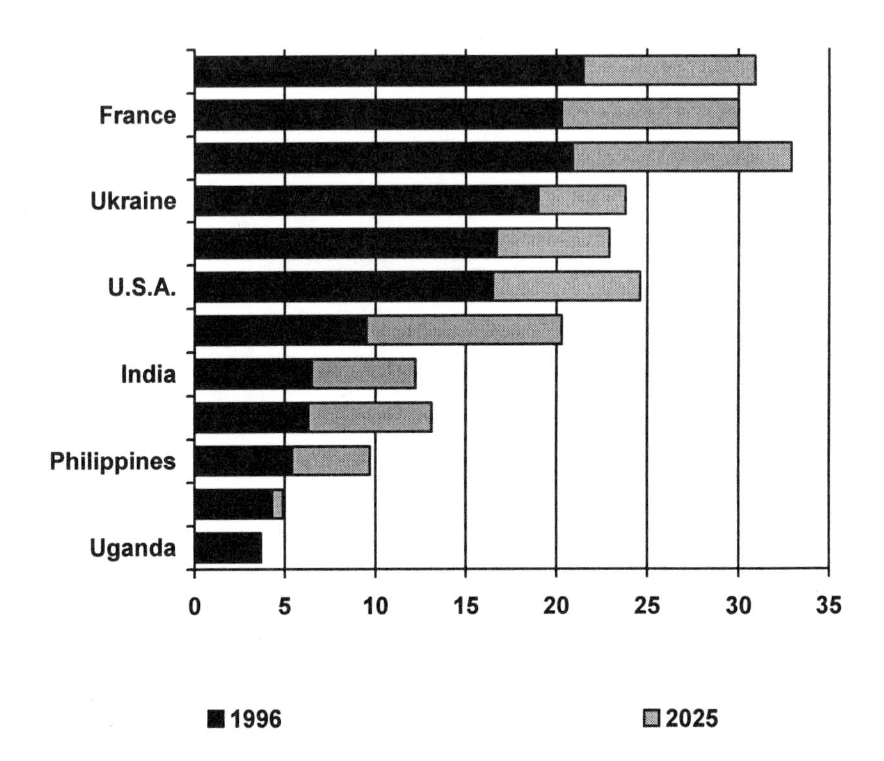

Source: U.S. Bureau of the Census, I.P.C., 1996

Figure 3. Percent of the Elderly in Various Countries, 1996, 2025

Many developing countries are at the other extreme, with 4-5 percent of the elderly among the entire population. The number of the world elderly is growing much faster than the entire population (Fig.4).

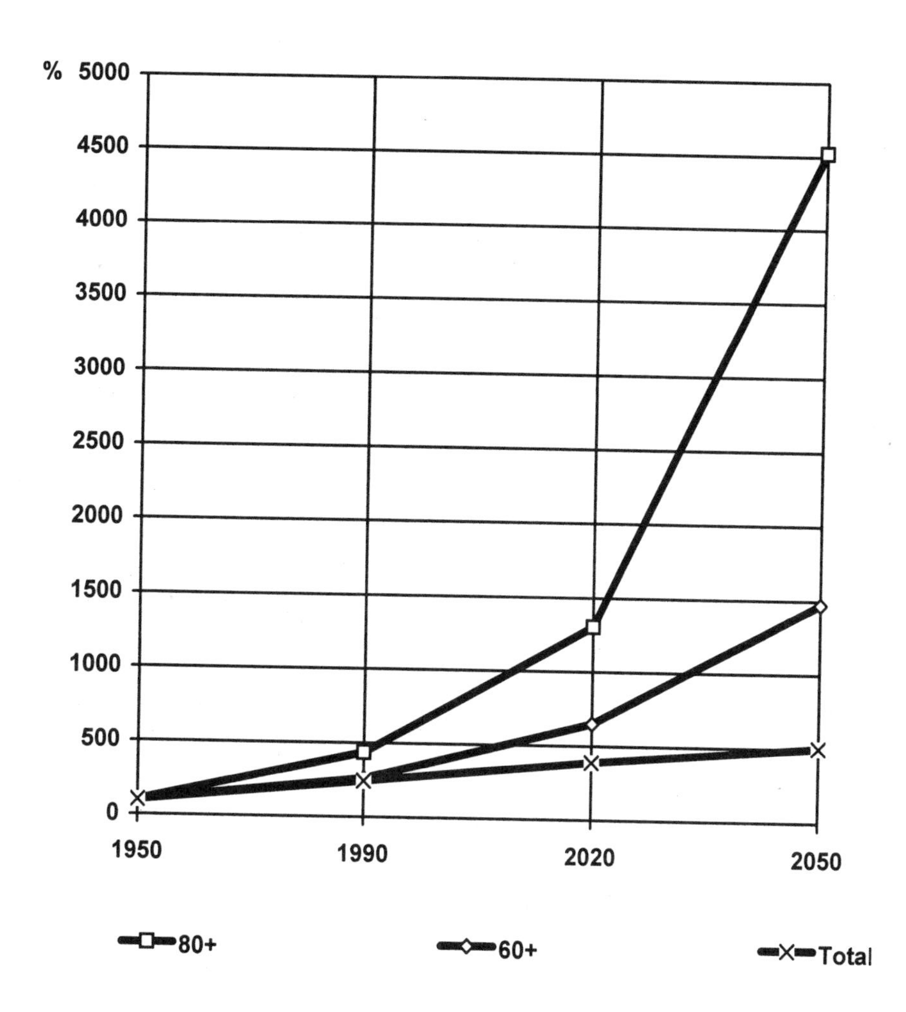

Source: United Nations, 1994

Figure 4. Population growth in the World 1950 - 2050

By the year 2020 more than 1 billion people in the world will be over 60 years old (in comparison with 550 million in 1996). By that time almost 300 million elderly will live in China, 165 million in India, more than 80 million in the USA, and almost 35 million in Russia.

The number of older people is increasing, both in developed and developing countries, but in the latter their growth occurs more intensively (Figs. 5 and 6). At present, almost two-thirds of the world elderly live in more developed countries, while in 20-25 years two-thirds of old people will live in developing countries.

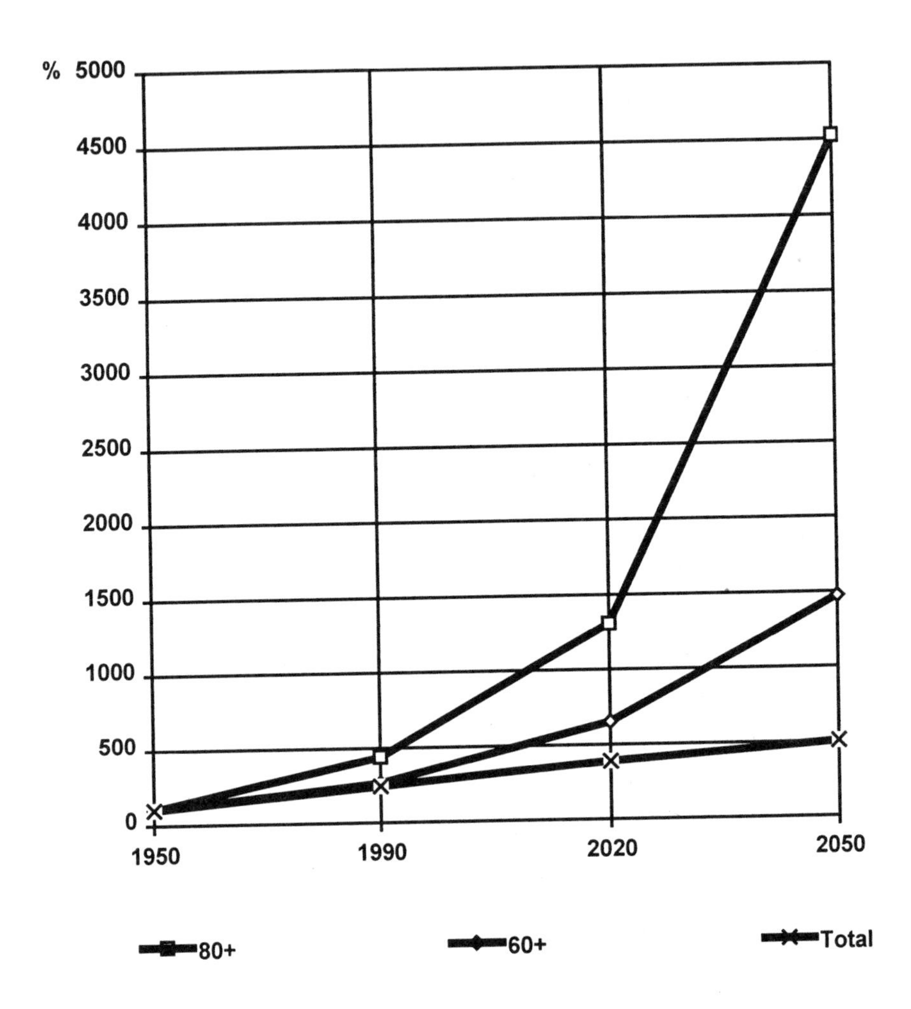

Source: United Nations, 1994

Figure 5. Population growth in more developed countries, 1950 - 2050

From 1995 to 2025 in Africa, the total population will double (from 730 million to 1.5 billion), while that of the elderly population will increase almost three-fold (from 36 million to 96 million (5). The same is true for the developing countries of Asia and Latin America.

The Oldest Old

The most dramatic increase has occurred in the numbers of the oldest old and centenarians, and it is projected that this population will continue to increase faster than the general population (see Figs. 4,5 and 6).

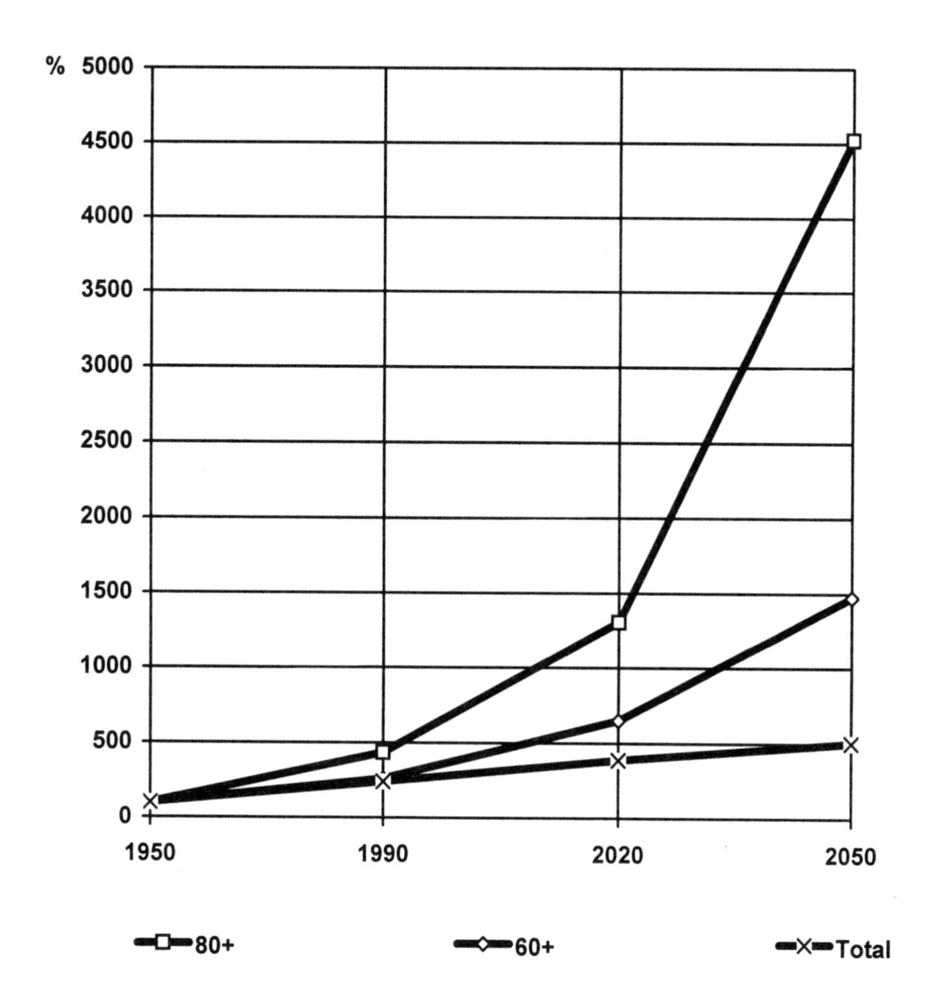

Source: United Nations, 1994

Figure 6. Population growth in less developed countries, 1950 - 2050

From 1997 to 2025, the world population will increase by 36 percent, the number of world elderly (aged 60 years and over) by 112 percent, and those aged 80 years and over by 152 percent (7). In 1996, the number of people aged 75 years and over increased by 1-2 percent in developing countries and up to 6-7 percent in developed countries. (The highest figure, 8.5 percent, was found in Sweden.) In the year 2025, this number will be surpassed in more than 20 countries.

The most fascinating aspect of longevity is the issue of centenarians, who exemplify the potential of human being to live 1.5 times longer than the average life expectancy worldwide. That is why various aspects of longevity were investigated in longitudinal and cross-sectional studies in the former Soviet Union, USA, Germany, Italy, Hungary, Sweden, Japan, China and Denmark (1,3,6,11,12,13,16,19 and 20).

Geographical distribution of centenarians is uneven. There are areas with higher concentrations of people who live to great old age. For example, in the USA it is Iowa and in the former Soviet Union the Caucasus region (Abkhasia, Azerbaijan). This phenomenon of group longevity can be found in other places, like in the Western region of Ukraine, or in Siberia or the Far East in Russia.

Statistics on centenarians is not consistent and in many countries is not available. Serious difficulties exist in registration and age validation of centenarians especially in societies with political or economic motivations to exaggerate age or numbers of centenarians. The data of population censuses compiled in the 1980s for 16 countries by Hungarian researchers show a significant variability of figures from country to country (3) (Table 1). When calculated per million, it differed more than 10 times, reaching 60-70 per 1 million.

During the last decades, their numbers have increased in most countries of the world, with rare exceptions (e.g., in some newly independent states of the former Soviet Union or in some other countries of Central and Eastern Europe, where socioeconomic upheavals, environmental pollution or military conflicts took place recently. For example, in the Ukraine in 1959 there were 2400 centenarians, in 1989 there were 2829, in 1996 only 1513.)

Table 1. Centenarians in Some Countries in 1980s

Country	Year	Persons 100 year and older	Per 1 mln population
Bulgaria	1984	1,792	199
Spain	1981	3,266	87
Ireland	1981	241	70
Portugal	1981	215	66
Norway	1984	245	59
New Zealand	1981	177	56
Japan	1980	2,000	17
Hungary	1980 1987	105 138 (96)	8 16 (11)

Source: Centenarians in Hungary, Karger 1990

FACTORS INFLUENCING LONGEVITY

Life expectancy and longevity are influenced to different extents by a variety of factors like heredity, gender, social environment, socioeconomic status, education, occupation, life styles, etc.

Researchers agree that one of the most important factors of longevity is heredity. Cases of familial longevity (genealogical studies), results of cytogenetic and molecular-genetic studies give strong support to this idea, expressed in a joke: "The best way to be a centenarian is to choose centenarian parents."

Another biological factor—gender—is also important. Females have a better chance to reach an advanced old age. In most countries, life expectancy for females is higher in comparison to males. Worldwide, the difference in life expectancy was 4 years in 1995 (67 years for females, and 63 years for males). In some countries, due to male supermortality in adult life, the difference in life expectancy for males and females are 11-12 years (Russia, Ukraine), while in some other countries the difference in life expectancy is minor or non-existent (India, Pakistan).

In 1995, 302 million of the world's elderly were women, while only 247 million were men. With advancing age, women are increasingly in the majority (Fig.7).

In centenarians, the males/females ratio is reaching 1:4 (Czech Republic and Slovakia, France, Hungary, Ukraine) or even 1:7 (Bulgaria, Finland).

It is believed that nutrition is of great importance in promoting longevity. Although traditional stereotypes of nutrition and diets of centenarians vary from country to country, many findings support the idea that they eat "healthier" diets. Their nutrition is characterized by moderate calorie-restricted diets, higher consumption of milk and dairy products and vegetables, food rich in vitamins and antioxidants, in thyrosine and linoleic acid and low in tryptophan (2,9).

At the last International Congress of Gerontology, cross-national comparisons were made on the personality of centenarians in France, Hungary, Japan, USA, Sweden. Although characteristics varied from culture to culture, it was concluded that emotional stability, low anxiety level, energy, sociability and commitment lead to high life satisfaction and good quality of life in centenarians and may be included in the predictive variables that determine longevity and well-being in the very old. The same was found in studies of the Kiev Institute of Gerontology in centenarians of Abkhasia and Azerbaijan. Social support, family status and familial support also contribute to longevity in various societies.

Complex clinical investigation of centenarians revealed the later onset and lower prevalence of age-related pathologies, like ischemic heart disease, cerebrovascular pathology, hypertension, parkinsonism and other forms of extrapyramidal insufficiency, as well as cancers.

The prevalence of cardiovascular diseases in Abkhasians and nervous diseases in Azerbaijanians was found in our study of the centenarians of the Caucasus. Correspondingly, changes in brain functions with aging were more intensive in Azerbaijanians, and those of the cardiovascular system in Abkhasians. From this we may conclude that functional mechanisms promoting longevity may vary from one ethnic group to another.

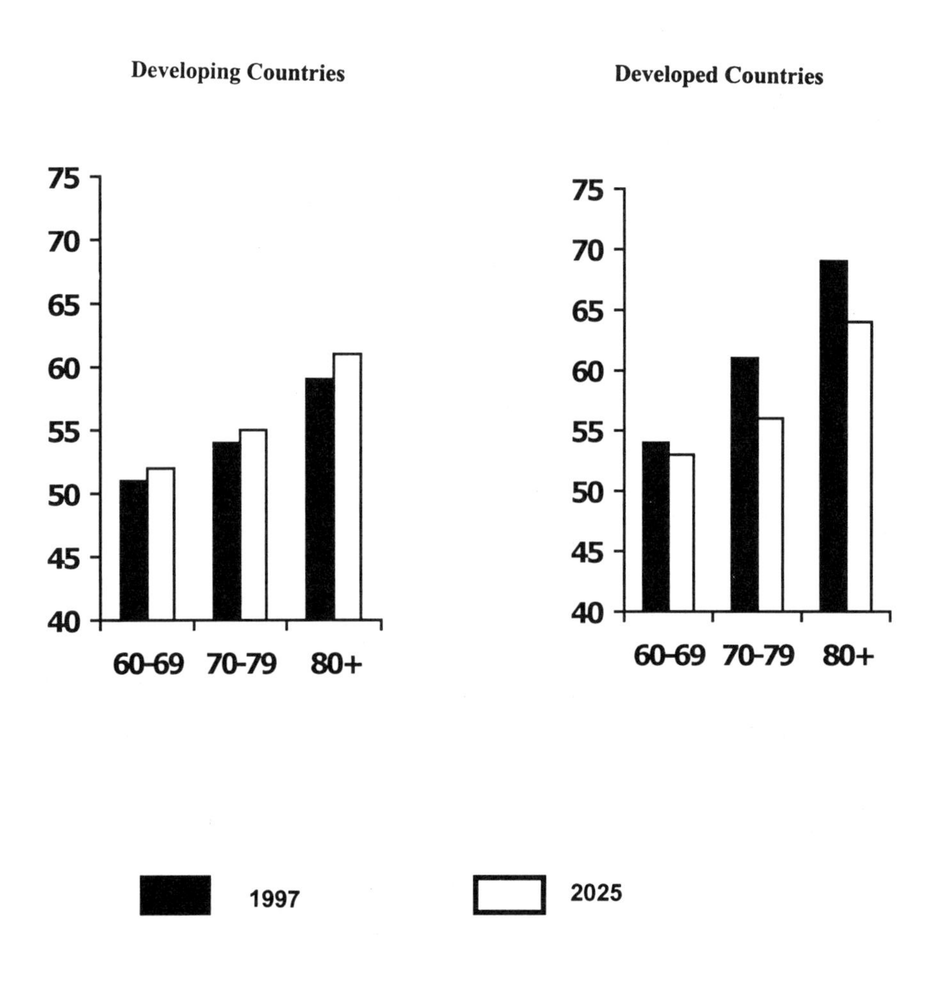

Source: U.S. Bureau of the Census, I.P.C., 1996

Figure 7. Percent Females in Older Age Groups: 1997 and 2025

Some data are consistent and some contradict each other, but there is agreement that although consistencies or inconsistencies exist in findings on the input of a specific factor, longevity is determined by a constellation of biological, psychological and social factors (Table 2).

Table 2. BOLSA Theoretical Model (Slightly Modified)

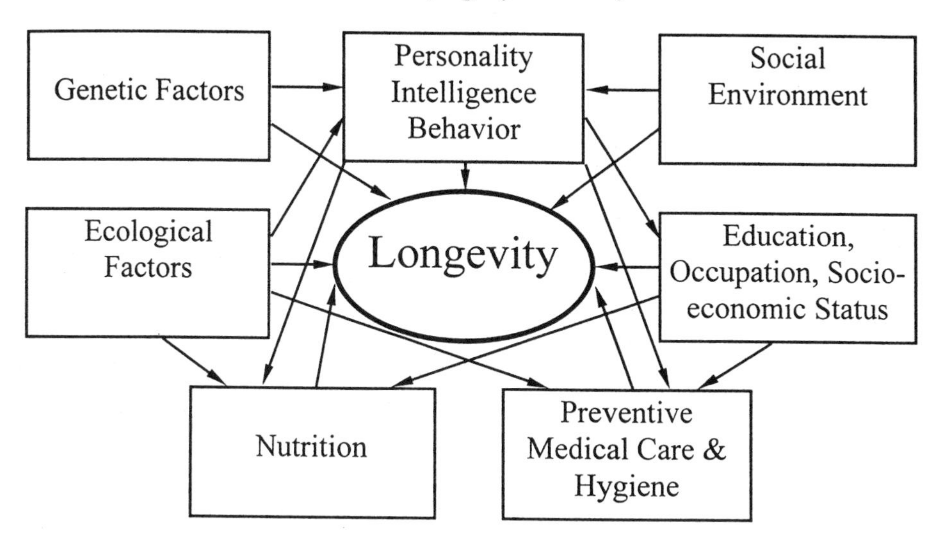

Life expectancy and longevity at a national level greatly depend on the economic development of the country: the higher indices of GNP and, especially, personal consumption, the higher life expectancy for the population (Table 3).

Table 3. Life Expectancy and Economy

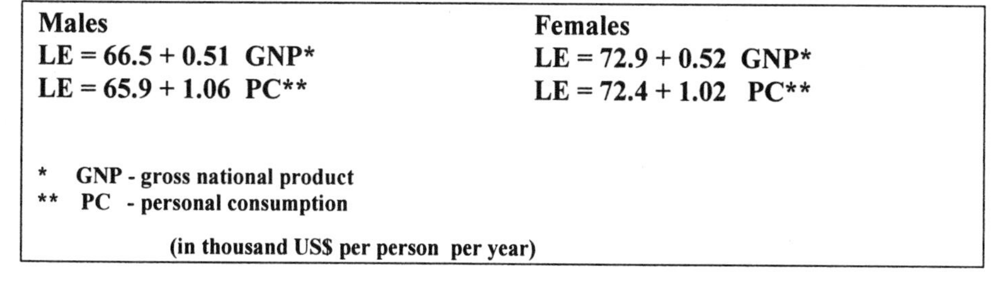

Males	**Females**
LE = 66.5 + 0.51 GNP*	**LE = 72.9 + 0.52 GNP***
LE = 65.9 + 1.06 PC**	**LE = 72.4 + 1.02 PC****

*** GNP - gross national product**
**** PC - personal consumption**

(in thousand US$ per person per year)

The future gains in longevity will depend on socioeconomic stability and world development, on progress in basic research in the biology of aging, on implementation of individual healthy life styles, and corresponding societal approaches and policies.

REFERENCES

1. Abkhasian Longevity. Editor-in-Chief: V.I. Kozlov. Nauka Publ., Moscow, 1987.

2. Babeanu S., Galaftion S., Nicolescu R. Food intake in promoting longevity. The Journal of Nutrition, Health & Aging 1997; 1(2)82.

3. Centenarians in Hungary. A Sociomedical and Demographic Study. Volume Editor: Edit Beregi. In: Interdisciplinary Topics in Gerontology. Vol. 27. Series Editor: H.P.von Hahn. Karger, Basel et al., 1990.

4. Demographic Yearbook, 1995. United Nations. N.Y., 1997.

5. D.Radha Devi. The aged in Africa: A situation analysis. Bold 1998; 8:2.

6. The Georgia Centenarians Study. Ed.: Leonard W. Poon. Intl Journal of Aging and Human Development 1992;34:1.

7. Gist Y.J. and Velkoff V.A. Demographic dimensions. International Brief Gender and Aging. U.S. Department of Commerce Economics and Statistics Administration. Bureau of the Census, December 1997.

8. Global Aging in the 21st Century. U.S. Department of Commerce Economics and Statistics Administration. U.S. Census Bureau, 1996.

9. Jeune B., Skytthe A. and J.W. Vaupel. The demography of centenarians in Denmark (in Danish). Ugeskr Laeger 1996 Dec 16;158(51):7392-7396.

10. Kuznetsova S.M. National and ethnic traditions and longevity. Age Nutrition 1996;7:26-29.

11. Lehr, U. and R. Schmitz-Scherzer. Psychosoziale Korrelate der Langlebigcheit. Acta Gerontologica 1974;4:261-268.

12. Maestroni, M.D. Centenarians in Milan. Ann Ital Med Int 1995 Jan; 10(1):46-48.

13. Phenomenon of Longevity. Anthropologic-Ethnographic Aspect. Editor-in-Chief: S.I.Brook. Nauka Publ., Moscow, 1982.

14. Pushkova, E.S., Klemina, L.V. and Ivanova, L.V. Longevity: Comparative analysis of the state of the problem in St. Petersburg and Iova State, USA. Klinicheskaya Gerontologiya (in Russian), 1997; 3:53-58. 13.

15. The Sex and Age Distribution of the World Population. The 1994 Revision. United Nations, N.Y., 1994.

16. Thomae H. Patterns of Aging: Findings from the Bonn Longitudinal Study of Aging. Karger, New York, 1976.

17. Torrey Barbara Boyle, Kevin Kinselia and Cynthia M. Taeuber. International Population Reports Series P-95, No. 78. An Aging World. September 1987.

18. Women, Aging and Health. Achieving health across the life span. World Health Organization, Geneva, 1996.

19. Wu Cangping. The Aging of Population in China. The International Institute on Ageing (INIA/CICRED), United Nations, Malta, 1991.

20. Zheng, Z.X., Wang, Z.S., Zhu, H.M., Yang, J.Y., Peng, H.Y., Wang, L.X, Li, J., Jiang, X.W., Yu, Y.F. Survey of 160 centenarians in Shanghai. Age Ageing 1993 Jan; 22(1):16-19.

LONGEVITY AS AN ARTIFACT OF CIVILIZATION

James W. Vaupel

Max Planck Institute for
Demographic Research
Doberaner Street 114
D-18057
Rostock, Germany

INTRODUCTION

The populations of most of the nations of the world are growing older. This shift is creating a new demography, a demography of low fertility and long lives. The rapidly growing aging populations are putting unprecedented stresses on societies, because new systems of financial support, social support, and health care have to be developed and implemented. A century ago most of the people born around the world died before they had children; most of the people who had children died before their children had children. Old people were unusual; extended, three-generation families represented only a small fraction of all families. Today, in developed countries and many developing countries as well, the typical newborn can expect to survive to see the birth not only of children and grandchildren but great-grandchildren as well.

I will touch briefly today on some of the health, social, and economic issues arising from the rapid growth in the numbers of older people and the shift of the age-distribution of populations to older age. I cannot even start, however, to do justice to all the interesting research demographers have done on the problems and opportunities associated with population aging. So I would like to focus my talk on a particular aspect of research, namely demographic analyses of survival and longevity.

RAPID GROWTH OF THE OLDER POPULATION

In countries where reliable data are available on centenarians, their numbers are increasing at an exceptionally rapid rate, about 8 percent per year on average. Demographers are used to population growth rates around 1 percent per year; an 8

Longevity and Quality of Life, Edited by Butler and Jasmin
Kluwer Academic / Plenum Publishers, New York, 2000.

percent growth rate seems more like an inflation rate. In England and Wales, an average of 74 persons per year reached age 100 between 1911 and 1920; by 1990 the number of people celebrating their 100th birthday had increased to almost 2000 and in 1997 the number will be around 3000 (Vaupel and Jeune 1995). Zeng Yi and I estimate that the number of centenarians in China is doubling every decade. In 1990 there were about 6000 people age 100 and above in China. By the year 2000 there may be more than 12,000. The population of centenarians is growing, in part, because of the increase in births a century ago, the sharp decline in infant and childhood mortality, and the substantial decline in mortality from childhood up to age 80. Demographic analysis demonstrates, however, that by far the most important factor in the explosion of the centenarian population—two or three times more important than all the other factors combined— has been the decline in mortality after age 80 (Vaupel and Jeune 1995).

Increases in maximum human longevity are also largely attributable to improvements in survival at the highest ages. Lundström (1995) carefully verified the ages of the oldest people who died in Sweden from 1860 through 1994. In the 30 years between 1860 and 1889, no one survived to age 106. Over successive decades, the maximum gradually rose, with the current Swedish record holder having died at age 112 in 1994. As argued by Jeune (1995), it is possible in Sweden (and other countries with modest populations) that no one attained the age of 100 before 1800.

There may have been a few scattered centenarians in earlier centuries, perhaps one per century somewhere in the world, perhaps even fewer (Wilmoth 1995). Zhao Zhongwei (1995) presents some evidence, from his study of the genealogy of the Wang family in China, of a man, Wang Xinglian, who may have died at age 102 in the year 1513. Tan Qihua, on the other hand, in research that has not yet been published, found no plausible examples of centenarians among 4362 famous Chinese who died before 1900 and who are listed in the *Dictionary of Historical Chinese Figures*. In contrast to the very rare sightings of plausible centenarians in past centuries, fully 100,000 genuine centenarians around the world may be alive to welcome the year 2000 (Vaupel 1994).

Wilmoth's (1995) analysis indicates that "there were almost certainly no true super-centenarians (individuals aged 110 or above) prior to the mortality decline of the past two or three centuries." Research by Peter Laslett and colleagues suggest that the first reasonably-well-documented case of a super-centenarian is Katherine Plunket, who died at the age of 111 in 1932 (Jeune 1995). Jeanne Calment is the first carefully-verified instance of a person reaching age 120 (Allard, Lebre, Robine 1994); she died at the age of 122 years and 5 months in August 1997.

Centenarians are still unusual and super-centenarians are a thousand-fold rarer, but these findings do illustrate the fact that mortality reduction can have a major impact on population growth at older ages and on extending the frontier of survival. The growth of the population of female octogenarians in England and Wales provides another telling example. The remaining life expectancy of 80-year-old females in England and Wales around 1950 was approximately 6 years. Currently, the corresponding figure is about 9 years, some 50 percent higher. As a result, the population of female octogenarians in England and Wales is roughly half again as big as it would have been if mortality *after age 80* had remained at 1950 levels. Putting this in terms of the population count, more than a half million females age 80+ are alive today in England and Wales who would have been dead if mortality after age 80 had not been reduced.*

Table 1 provides information about the size of the older population of various countries, from age 60 and up, for both sexes combined. Estimates are also given for the size of these populations in 2025. The projections assume slow improvements in mortality, so I believe that the estimates for 2025 are likely to be low. Nonetheless, the size of the older population shows substantial increases, not only in Europe but in Japan, the U.S., China, and India as well.

Table 1. Proportion of Population above Age 60 (in %) and Population above Age 60 (in millions) for Selected Countries in 1996 and Projected for 2025.

	% 60+		Millions 60+	
	1996	2025	1996	2025
Italy	22	33	13	18
Japan	21	33	26	40
Germany	21	32	7	28
France	20	30	12	18
U.K.	21	29	12	17
U.S.A.	17	25	44	83
China	9	20	115	290
Brazil	7	16	11	31
Mexico	7	13	6	18
India	7	12	62	165
South Africa	7	10	3	6
Egypt	6	10	4	10

Source: U.S. Bureau of the Census (1997).

Zeng Yi, Wang Zhenglian and I have recently completed some research on the growth of the old and very old populations in China, up to the year 2050 (Zeng, Vaupel, Wang 1997). Under our low-mortality scenario, which I think may capture the decline in death rates in the future, we estimate that the 65+ year-old population of China will grow from 63 million in 1990 to more than 400 million in 2050, a 6-fold increase. For the 85+ population, we estimate growth from 2.3 million in 1990 to more than 80 million in 2050, an extraordinary 36-fold explosion. By the middle of the next century, then, the oldest-old population, age 85 and older in China may exceed the total population of Germany.

IMPROVEMENTS IN SURVIVAL

Why is the frontier of survival advancing to higher and higher ages? Why are the numbers of the older population growing so rapidly? The answer is that longevity is an artifact of civilization. It is often stated that there is a fixed maximum human life span and this span is determined by our genetic makeup. It is claimed that death rates at advanced ages are the same today as they were in ancient and even prehistoric times (Fries and Crapo 1981). The truth is that death rates at even the most advanced ages are plastic and can be reduced by environmental, behavioral, and medical interventions. There is no evidence that there is a fixed maximum life span. Instead, there is compelling evidence that over the past 50 years remarkable improvements have been made in increasing survival at older ages.

Despite the evidence, many people still believe that old-age mortality is intractable. Because of its implications for social, health, and research policy, the belief is pernicious; forecasts of the growth of the elderly population, expenditures on life-saving health care for the elderly, and expenditures for biomedical research on the deadly illnesses of old age are all too low.

Far from being intractable, mortality at older ages has fallen dramatically since 1950 in developed countries and most developing countries as well. Tables 2 and 3 present a few figures, but extensive documentation is available elsewhere (Kannisto 1994, 1996; Kannisto et al. 1994; Vaupel et al. 1998).

Table 2 shows the rates of improvement in female mortality at older ages in the Nordic countries of Denmark, Finland, Norway, and Sweden. These countries have exceptionally reliable data up to the highest ages. The rate of improvement is impressive, especially since the 1960s. Also striking is the acceleration of mortality improvement over time. Male death rates are higher than female rates at all ages from conception to advanced old age. Male rates of mortality improvement have been slower than female rates. But substantial gains have also been made for males in recent decades.

Table 2. Average annual rate of improvement in female mortality (in %) for aggregation of Denmark, Finland, Norway and Sweden, for sexagenarians, septuagenarians, octogenarians, and nonagenarians, over successive 20-year periods.

	Age Category			
Time Period	**60s**	**70s**	**80s**	**90s**
1900s-1920s	0.3	0.2	0.1	0.0
1920s-1940s	0.7	0.4	0.2	0.0
1940s-1960s	1.7	1.0	0.6	0.5
1960s-1980s	1.5	2.1	1.7	1.2

Source: Own calculations. See Kannisto et al. (1994) for description of data and of how average annual rates of improvement are calculated.

Table 3 displays death rates by age and time for females in the Nordic countries. The increase in death rates with age is dramatic. The decrease in death rates over time is also noteworthy. If mortality is reduced, then the number of lives saved is proportional to the absolute decline rather than the relative decline. For instance, if the probability of death at a particular age is reduced from 20 percent to 15 percent, then an extra 5 percent of the population continue to enjoy life. In the last row of Table 3, the absolute improvements in Nordic female mortality are displayed. It is at the most advanced ages that the most lifesaving has occurred. On second thought, this may not seem so surprising because it is at the highest ages that death rates are very high. Moreover, lives saved at the highest ages are generally not extended for more than a few years. Nonetheless, the large absolute reductions in mortality among centenarians and nonagenarians is a remarkable achievement, at sharp variance with the view that old-age mortality is intractable.

Table 3. Female central death rates (in %) for aggregation of Denmark, Finland, Norway and Sweden for sexagenarians, septuagenarians, octogenarians, nonagenarians, and centenarians, in two periods, 1930-49 and 1989-1993.

Age:	**60s**	**70s**	**80s**	**90s**	**100s**
Time Period					
1930-49	2.4	6.4	16.1	33.9	70.1
1989-93	1.1	3.1	9.1	23.4	48.5
Variance	1.3	3.3	7.0	10.5	21.6

Source: Own calculations. See Kannisto et al. (1994) for description of data and of how average annual rates of improvement are calculated.

VARIATION IN LIFE SPAN

The multiplication of the population of older people heightens interest in a fundamental question: Why do some people die at 60, others at 80, and a few at 100? Why are the odds of dying at 80 rather than 60 increasing and the chance of surviving to 100 rapidly increasing (albeit from a very low level)? How important are genetic versus environmental, behavioral, and medical factors in determining how long an individual will live?

It might be expected that the answers to these questions—and the determinants of longevity more generally—are well understood. The duration of life has captured the attention of people for thousands of years. The life span for humans as well as for other species can be readily measured. Many countries have a huge array of statistical data going back for many centuries.

A recent review, however, of the determinants of longevity (Christensen and Vaupel 1996) concludes that surprisingly little is known. The chance of reaching age 80 (or 90 or 100) is better for:

- women than men
- people born in this century rather than earlier
- people born in developed countries, and
- people who have favorable genes, such as the ApoE 2 gene (Schächter et al. 1994).

Smoking is certainly a health hazard for young and old. Obesity may be a risk factor and diet is probably important. Some pharmaceuticals, such as DHEA, may increase survival at older ages. Studies of twins and other related individuals suggest that about 25 percent of the variation in adult life expectancy appears to be attributable to genetic variation among individuals (McGue et al. 1993, Herskind et al. 1996). Research in progress by Anatoli Yashin and Ivan Iachine suggests that an additional 25 percent may be attributable to non-genetic characteristics that are more or less fixed by the time a person is about 30: characteristics such as educational achievement, socioeconomic status, parents' ages at a person's birth, etc. Research on the relative importance for longevity of various candidate genes and non-genetic fixed attributes is, however, still at an early stage of development.

Barker's (1992, 1995) "fetal-origins hypothesis" suggests that nourishment *in utero* and during infancy, programs the development of risk factors for several important diseases of middle and old age. Other researchers have also concluded that nutrition and infections early in life have major effects on adult mortality (Kermack, McKendrick, McKinlay 1934, Elo and Preston 1992, Fogel 1993). To the extent this is true, longevity may be determined by conditions in childhood and, perhaps, even before birth. There is, however, conflicting evidence to suggest that current conditions (i.e., at older ages) may be much more important than conditions early in life. Kannisto (1994, 1996) finds period effects to be much more important than cohort effects on mortality after age 80. Christensen et al. (1995) find that from age 6 up to the oldest ages, twins (who tend to be born prematurely and at low birth weight) suffer the same age-specific death rates as singletons. And Kannisto, Christensen, and Vaupel (1997) find "no increased mortality in later life for cohorts born during famine." Pinning down the nature and magnitude of possible lingering effects of early-life conditions on survival at advanced ages is an important research priority.

CONCLUSION

Over the past half century and especially in the most recent decades, remarkable improvements have been achieved in survival at older ages, especially at the highest ages. This progress has accelerated the growth of the population of older people and has advanced the frontier of human survival substantially beyond the extremes of longevity attained in pre-industrial times.

Little, however, is yet known about why mortality among the oldest-old has been so plastic since 1950. There is considerable (but still inadequate) knowledge of why some people die in infancy or childhood and why some people die prematurely at adult ages before age 60 or 70. Much less is known about why some people survive to age 80, others to age 90, and a few to age 100. The little that is known has largely been learned within the past few years and new findings (especially concerning genetic factors) are emerging at a rapid rate.

Although we do not yet know the causes of the remarkable improvements in old-age survival, we do know that they have occurred and that mortality, even at advanced ages, is highly changeable. The intractability of mortality at older ages is untenable. Genetic factors play some role in how long a person lives; non-genetic factors play an even more important role. The progress made since 1950 in reducing mortality at older ages is entirely attributable to non-genetic changes. Longevity is an artifact of civilization, one of the towering achievements of modern life.

Endnote

* The figure of 9 years for the remaining life expectancy of 80-year-old females in England and Wales is an estimate for 1997, based on data from Kannisto (1996). It is not a precise figure, but good enough for illustrative purposes. Change in life expectancy is often (but not always) serviceable as a rough indicator of the impact of mortality reductions on population size; the required calculations to produce a more exact estimate are fairly complicated. See Kannisto (1996) or Vaupel and Jeune (1995) for details. Kannisto's calculations indicate that mortality improvements after age 80 in England and Wales between 1960 and 1990 (rather than 1950 and 1997) increased the female population by 250,000 persons.

ACKNOWLEDGMENTS

Much of the research reported is the result of collaborative work with James R. Carey, Kaare Christensen, James W. Curtsinger, Bernard Jeune, Vaino Kannisto, Peter Laslett, Hans Lundström, Kenneth G. Manton, Matt McGue, Cindy R. Owens, A. Roger Thatcher, Anatoli I. Yashin, and Zeng Yi. Some of the research was supported by the U.S. National Institute on Aging through grant NIA-08761. Other support was provided by the Danish Research Council and by the Max Planck Society.

Parts of this chapter were previously published in my article, "The remarkable improvements in survival at older ages," Phil. Trans. Roy. Soc. Lond B.352 (1997), pages 1799-1804.

REFERENCES

1. Allard, M., V. Lebre, J-M. Robine (1994) "Les 120 ans de Jeanne Calment," Le Cherche Midi Editeur, Paris.

2. Barker, D.J.P. (1992) Fetal and Infant Origins of Adult Disease, British Medical Journal Press, London.

3. Barker, D.J.P. (1995) "Fetal Origins of Coronary Heart Disease," British Medical Journal 311:171-4.

4. Christensen, K., J.W.Vaupel, N.V. Holm, A.I. Yashin (1995) "Mortality among Twins after Age 6: Fetal Origins Hypothesis versus Twin Method," British Medical Journal 310: 432-6.

5. Christensen, K., J.W. Vaupel (1996) "Determinants of Longevity: Genetic, Environmental, and Medical Factors," Journal of Internal Medicine.

6. Elo, I.T., S.H. Preston (1992) "Effects of Early-Life Condition on Adult Mortality: A Review," Population Index 58(2):186-222.

7. Fogel, R.W. (1993) "Economic Growth, Population Theory, and Physiology: The Bearing of Long-Term Processes on the Making of Economic Policy," The Nobel Foundation, Stockholm, Sweden.

8. Fries, J.F., L.M. Crapo (1981) Vitality and Aging, W.H. Freeman, San Francisco.

9. Herskind, A.M., M. McGue, N.V. Holm, T.I.A. Soerensen, B. Harvald, and J.W. Vaupel (1996) "The Heritability of Human Longevity," Human Genetics 97:319-323.

10. Jeune, B., J.W. Vaupel, editors, (1995) Exceptional Longevity: From Prehistory to the Present, Odense University Press, Denmark.

11. Jeune, B. (1995) "In Search of the First Centenarians" in Jeune and Vaupel (1995).

12. Kannisto, V. (1994) Development of Oldest-Old Mortality, 1950-1990, Odense University Press, Denmark.

13. Kannisto, V. (1996) The Advancing Frontier of Survival: Life Tables for Old Age, Odense University Press, Denmark.

14. Kannisto, V., K. Christensen, J.W. Vaupel (1997) "No Increased Mortality in Later Life for Cohorts Born during Famine," American Journal of Epidemiology 145(11):987-994.

15. Kannisto, V., J. Lauritsen, A. R. Thatcher, J.W. Vaupel (1994) "Reductions in Mortality at Advanced Ages," Population and Development Review 20(4):793-810.

16. Kermack, W., A. McKendrick, P. McKinlay (1934) "Death-Rates in Great Britain and Sweden: Some General Regularities and Their Significance," The Lancet 1:698-703.

17. Lundström, H. (1995) "Record Longevity in Swedish Cohorts Born since 1700" in B. Jeune, J.W. Vaupel (1995).

18. McGue, M., J.W. Vaupel, N. Holm, B. Harvald (1993) "Longevity Is Moderately Heritable in a Sample of Danish Twins Born 1870-1880", Journal of Gerontology 48:B237-B244.

19. Schächter, F., L. Faure-Delanef, F. Guenot, H. Rouger, P. Forguel, L. Lesueur-Ginot (1994) "Genetic Associations with Human Longevity at the APOE and ACE Loci," Nature Genetics 6:29-32.

20. Thatcher, A.R. (1992) "Trends in Numbers and Mortality at High Ages in England and Wales," Population Studies 46:411-426.

21. Thatcher, A.R., V. Kannisto, J.W. Vaupel (1997) The Force of Mortality at Ages 80 to 120, Odense University Press, Denmark.

22. Bureau of the Census (1997) "Global Aging into the 21st Century," Washington, D.C..

23. Vaupel, J.W. (1994) Preface to Kannisto (1994).

24. Vaupel, J.W. (1977) "Trajectories of Mortality at Advanced Ages" in K.W. Wachter and C.E. Finch (eds.) Between Zeus and the Salmon: The Biodemography of Longevity, National Academy Press, Washington, D.C.

25. Vaupel, J.W., B. Jeune (1995) "The Emergence and Proliferation of Centenarians" in Jeune and Vaupel (1995).

26. Vaupel, J.W., H. Lundström (1994) "The Future of Mortality at Older Ages in Developed Countries" in W. Lutz (ed.) The Future of the Population of the World, Earthscan, London.

27. Vaupel, J.W., Z. Wang, Y. Zeng (forthcoming) "Han Chinese Mortality at Older Ages."

28. Wilmoth, J. (1995) "The Earliest Centenarians: A Statistical Analysis" in. Jeune and Vaupel (1995).

29. Wilmoth, J.R. (1997) "In Search of Limits: What Do Demographic Trends Suggest about the Future of Human Longevity" in K.W. Wachter and C.E. Finch (eds.) Between Zeus and the Salmon: The Biodemography of Longevity, National Academy Press, Washington, D.C.

30. Zeng, Y., J.W. Vaupel, Z. Wang (1997) "Household Projection Using Conventional Demographic Data" in W. Lutz and J.W. Vaupel (eds.) Rethinking Population Projections, forthcoming.

31. Zhao, Z. (1995) "Record Longevity in Chinese History" in Jeune and Vaupel (1995).

CAN WE AFFORD LONGEVITY?

Robert W. Fogel

University of Chicago
Graduate School of Business
1101 East 58th Street
Chicago, IL 60637
USA

INTRODUCTION

The nations of the Organization of Economic Cooperation and Development (OECD) generally are faced with crises in their pension and health care systems, not because they are poor but because they are, by historical or Third World standards, exceedingly rich. It is the enormous increase in their per capita incomes over the past century that permitted the average length of retirement to increase by five-fold, the proportion of a cohort that lives to retire to increase by seven-fold and the amount of leisure time available to those still in the labor force to increase by nearly four-fold (Appendices 5.2 and 5.3 in Fogel 1999; cf. Ausubel & Grübler 1995; Costa 1998; Lee 1996).

TECHNOPHYSIO EVOLUTION

Study of the long-term causes of these developments, particularly of the reduction in mortality, points to the existence of a synergism between technological and physiological improvements that has produced a form of human evolution that is biological but not genetic. It is rapid, culturally transmitted, and not necessarily stable. This process is still ongoing in both rich and developing countries, and is called "technophysio evolution."

Unlike the genetic theory of evolution through natural selection, which applies to the whole history of life on earth, technophysio evolution applies only to the last 300 years of *human* history, and particularly to the last century. Despite its limited scope technophysio evolution appears to be relevant to forecasting likely trends over the next century or so in longevity, the age of onset of chronic diseases, body size, and the efficiency and durability of vital organ systems (Fogel and Costa

Longevity and Quality of Life, Edited by Butler and Jasmin
Kluwer Academic / Plenum Publishers, New York, 2000.

1997). It also has a bearing on such pressing issues of public policy as the growth in population, in pension costs, and in health care costs.The theory of technophysio evolution rests on the proposition that during the last 300 years, particularly during the last century, human beings have gained an unprecedented degree of control over their environment—a degree of control so great that it sets them apart not only from all other species, but also from all previous generations of *Homo sapiens*. This new degree of control has enabled *Homo sapiens* to increase its average body size by over 50 percent, to increase its average longevity by more than 100 percent, and to improve greatly the robustness and capacity of vital organ systems.

Table 1
A Comparison of Energy Available for Work Daily per Consuming Unit in France, England and Wales, and the United States, 1700-1980 (in kcal)

Year	France (1)	England and Wales (2)	United States (3)
1700 (a)		720	2,313
1705	439		
1750		812	
1785	600		
1800		858	
1803-12			
1840			1,810
1845-54			
1850		1,014	
1870	1,671		
1880			2,709
1944			2,282
1975	2,136		
1980		1,793	1,956

(a) pre-Revolutionary Virginia

Source: Fogel & Floud, 1999

During its first 100,000 or so years, *Homo sapiens* increased at an exceedingly slow rate. The discovery of agriculture about 11,000 years ago broke the tight constraint on the food supply imposed by a hunting and gathering technology, making it possible to release between 10 and 20 percent of the labor force from the direct production of food, and also giving rise to the first cities. The new technology of food production was so superior to the old technology that it was possible to support a much higher rate of population increase than had existed prior to ca. 9000 BC. Yet, the advances in the technology of food production after the second Agricultural Revolution (which began about 1700 AD) were far more dramatic than the earlier breakthrough, since they permitted the population to increase at so high a rate that a line showing population growth would rise almost vertically. Abetted by a remarkable acceleration in the rate of technological change, the increase in world population between 1900 and 1990 was four times as great as the increase during the whole previous history of humankind.

The most important aspect of technophysio evolution is the continuing conquest of chronic malnutrition due mainly to a severe deficiency in dietary energy, which was virtually universal three centuries ago. Table 1 shows that in rich countries today some 1800 to 2000 kcal of energy are available for work by a typical adult male, aged 20-39. During the eighteenth century, however, France produced less than one-third the current U.S. amount of energy for work. And England was not much better off. Only the U.S. provided potential energy for work equal to or greater than late twentieth century levels during the eighteenth and early nineteenth centuries, although some of that energy was wasted due to the prevalence of diarrhea and other conditions that undermined the body's capacity to utilize nutrients. One implication of these estimates of caloric availability is that mature European adults of the eighteenth and much of the nineteenth century must have been very small and less active by current standards.

Recent studies have established the predictive power of height and weight at early ages with respect to onset of diseases and premature mortality at middle and late ages. Variations in height and weight are associated with variations in the chemical composition of the tissues that make up vital organs, in the quality of the electrical transmission across membranes, and in the functioning of the endocrine system and other vital systems. Nutritional status thus appears to be a critical link connecting improvements in technology to improvements in human physiology.

Research on this question is developing rapidly, and some of the new findings are yet to be confirmed. The exact mechanisms by which malnutrition and trauma *in utero* or early childhood are transformed into organ dysfunction are still unclear. What is agreed upon is that the basic structure of most organs is laid down early, and it is reasonable to infer that poorly developed organs may break down earlier than well-developed ones. The principal evidence so far is statistical and, despite agreement on specific dysfunction, there is no generally accepted theory of cellular aging (Tanner 1990, 1993).

With these caveats in mind, recent research bearing on the connection between malnutrition and body size and the later onset of chronic disease can conveniently be divided into three categories. The first category involves forms of malnutrition (including the ingestion of toxic substances) that cause permanent, promptly visible physiological damage, as is seen in the impairment of the nervous systems of fetuses due to excessive smoking or consumption of alcohol by pregnant women, or by deficiencies in folate, iron and iodine (Scrimshaw & Gordon 1968;

Martorell et al. 1990; Lozoff et al. 1991; Czeizel & Dudás 1992; Rosenberg 1992; Scrimshaw 1993; Chavez et al. 1994).

Not all damage due to retarded development *in utero* or infancy caused by malnutrition shows up immediately. Recent studies in England, Sweden, India, and elsewhere have reported that such conditions as coronary heart disease, hypertension, stroke, non-insulin dependent diabetes and autoimmune thyroiditis begin *in utero* or in infancy, but do not become apparent until mid-adult or later ages (Barker 1992, 1994). In these cases, individuals appear to be in good health, and function well in the interim. However, early onset of the degenerative diseases of old age appears to be linked to inadequate cellular development early in life.

Certain physiological conditions incurred by persons suffering from malnutrition can, in principle, be reversed by improved dietary intake, but they often persist because the causes of the malnutrition persist. If the malnutrition persists long enough these conditions can become irreversible or fatal; they include the degradation of tissue structure, especially in such vital organs as the lungs, the heart and the gastrointestinal tract. In the case of the gastrointestinal system, atrophy of the mucosal cells and intestinal *villi* result in decreased absorption of nutrients.

So far I have focused on the contribution of technological change to physiological improvements. The process has been synergistic, however, with improvement in nutrition and physiology contributing significantly to the process of economic growth and technological progress along the lines that I have described elsewhere. Here I merely want to point out the main conclusion. Technophysio evolution appears to account for about half of British economic growth over the past two centuries. Much of this gain was due to the improvement in human thermodynamic efficiency. The rate of converting human energy input into work output appears to have increased by about 50 percent since 1790 (1994).

The Virtue of Increased Spending on Health Care and Retirement

One aspect of technophysio evolution has been a change in the structure of consumption and in the division of discretionary time between work and leisure. Table 2 shows how sharply the U.S. distribution of consumption has changed over the past 119 years. Food, clothing and shelter, which accounted for 75 percent of consumption (expanded to include the value of leisure time) in 1875, accounted for under 20 percent of expanded consumption in 1994. Leisure, on the other hand, has risen from 15 percent of consumption to 64 percent. The expenditure category "other," which consists mainly of services, has risen from 10 to 20 percent of consumption, with the greatest increase in the category of health care (Fogel 1999, ch. 5 and Appendix 5.3). As Table 2 shows, the long-term income elasticities (the percent of change in consumption caused by a one percent change in income) of the demand for food and clothing are below 0.5 and the elasticity of the demand for shelter is close to but below 1.0. On the other hand, the elasticities of demand for leisure and for medical services are well over 1.

Even if the elasticities of these last two decline somewhat during the next 45 years, it is still likely that the burden of health care and retirement will require a substantially larger share of GDP than has been allowed in most recent forecasts. More conventional forecasts have been compromised, not only by the failure to take adequate account of the growth in the consumption of leisure, but by a failure to recognize differences in the time preference for leisure. Leisure can be taken while still in the labor force by varying daily, weekly, and annual hours of work. Or leisure can be taken in a block later in life by varying the age of retirement. Different households have different time preferences for leisure, which depend partly on the uses of leisure, and partly on their lifetime targets for levels of various categories of material consumption. On average, the consumption of leisure time after retirement has grown more rapidly than the consumption of leisure time before retirement. In 1880 only a quarter of life-time leisure was taken after retirement. Today nearly half is taken after retirement.

Table 2
The Long-Term Trend in the Structure of Expanded Consumption and the Implied Income Elasticities of Several Consumption Categories

Consumption Class	Distribution of Expanded Consumption (%) 1875	1994	Long-Term Income Elasticities
Food	49	5	0.20
Clothing	12	2	0.43
Shelter	14	8	0.84
Other	10	20	1.33
Leisure	15	64	1.61

Note: Expanded consumption is the sum of conventional expenditures plus the imputed value of leisure time. The income elasticity of demand is the percentage of increase in expenditures on a given commodity associated with a one percent increase in income. If the elasticity is less than one, it is said to be inelastic since consumption will increase more slowly than income.

Source: Fogel 1999.

What then is the virtue of increasing spending on retirement and health rather than on goods? It is the virtue of providing consumers in rich countries with what they want most. It is the virtue of not insisting that individuals must increase earn-work an extra ten hours a week or an extra 40,000 hours per lifetime in order to produce more food or durable goods than they want, just because such consumption will keep factories humming. The point is that leisure time activities (including lifelong learning) and health care are the growth industries of the late twentieth and the early twenty-first centuries. *They* will spark economic expansion during our age, just as agriculture did in the eighteenth and early nineteenth centuries, and as manufacturing, transportation and utilities did in the late nineteenth and much of the twentieth centuries.

The growing demand for health care services is not due primarily to a distortion of the price system but to the increasing effectiveness of medical intervention (Fogel 1999, Appendix 5.3). That increase since 1910 is strikingly demonstrated by comparing the second and last columns of the line on hernias in Table 3. Prior to World War II, once hernias occurred, they were generally permanent and often exceedingly painful conditions. However, by the 1980s about three-quarters of all veterans who ever had hernias were cured. Similar progress over the seven decades is indicated by the line on genito-urinary conditions, which shows that three-quarters of those who ever had such conditions were cured. Other areas where medical intervention has been highly effective include control of hypertension and reduction in the incidence of stroke, surgical removal of osteoarthritis, replacement of knee and hip joints, curing of cataracts, and chemotherapy that reduces the incidence of osteoporosis and heart disease (Manton 1993; Manton et al. 1997). It is the success of medical interventions combined with rising incomes that has led to a huge increase in the demand for medical services.

THE CAPACITY TO AFFORD ABUNDANT LEISURE

The forecasts based on the data reported in Table 2 imply that by 2040 those still in the labor force, as conventionally defined, would have over 50 hours per week of leisure, that the average age of retirement (the beginning of full-time leisure or the end of regular work) would begin about age 55, and that the average duration of full-time leisure would be about 35 years. Will OECD nations have the resources to afford amounts of leisure that would once have been considered luxurious and also provide high quality health care for an additional seven or eight years of life?

Assuming that the per capita income of OECD nations will continue to grow at a rate of 1.7 percent per annum, the resources to finance such expanded demands will be abundant (Fogel 1999, Appendix 5.4). This is a reasonable growth rate, below the long-term experience of the last half century, as well as of the past decade and a half (Maddison 1991). Consider a typical new American household established in 1995 with the head aged 20 and the spouse earning 36 percent of the income of the head (i.e., the spouse works part time) (Fogel 1999). Such a household could accumulate the savings necessary to retire at age 55, with a pension paying 60 percent of its peak life-cycle earnings, by putting aside 14.7 percent of annual earnings from the year that the head and spouse enter the labor force. That pension would permit retirees at age 55 to maintain their pre-retirement standard of living, with a real income that would rank them among the richest quarter of householders today (Fogel 1999).

By putting aside an additional 9.4 percent of income the household can buy high quality medical insurance that will cover the entire family until the children (two) enter the labor force, and also cover the parents' medical needs between the time they retire and age 83 (assumed to be the average age of death in their cohort). Saving an additional 7.8 percent of income will permit parents to finance the education of their children for 16 years, through the bachelor's degree at a good university.

Table 3

Comparison of the Prevalence of Chronic Conditions among Union Army Veterans in 1910, Veterans in 1983 (Reporting Whether They Ever Had Specific Chronic Conditions), and Veterans in U.S. National Health Interview Survey (NHIS) 1985-1988 (Reporting Whether They Had Specific Chronic Conditions during the Preceding 12 Months), Aged 65 and Above, Percentages

Disorder	1910 Union Army veterans	1983 veterans	Age-adjusted 1983 veterans	NHIS 1985-1988 veterans
Musculoskeletal	67.7	47.9	47.2	42.5
Digestive	84.0	49.0	48.9	18.0
Hernia	34.5	27.3	26.7	6.6
Diarrhea	31.9	3.7	4.2	1.4
Genito-urinary	27.3	36.3	32.3	8.9
Central nervous, endocrine, metabolic, or blood	24.2	29.9	29.1	12.6
Circulatory (a)	90.1	42.9	39.9	40.0
Heart	76.0	38.5	39.9	26.6
Varicose veins	38.5	8.7	8.3	5.3
Hemorrhoids (b)	44.4			7.2
Respiratory	42.2	29.8	28.1	26.5

(a) Among veterans in 1983, the prevalence of all types of circulatory diseases will be underestimated because of inderreporting of hemorrhoids.
(b) The variable indicating whether the 1983 veterans ever had hemorrhoids is unreliable

Notes: Prevailing rates of Union Army veterans are based on examinations by physicians. Those for the 1980s are based on self-reporting. Comparison of the NHIS rates with those obtained from physicians' examinations in National Health and Nutrition Examination Survey II (NHANES II) indicates that the use of self-reported health conditions does not introduce a significant bias into the comparison. See Fogel et al. (1993) for a more detailed discussion of possible biases and their magnitudes.

Source: Fogel et al. 1993.

What I have described is a provident fund of the type recently introduced or under consideration in some high-performing Asian economies and elsewhere (Iyer 1993; Poortvliet and Laine 1995). I have assumed that the savings would be invested in conservatively run funds, such as Teachers' Insurance and Annuity Association (TIAA)/College Retirement Equities Fund (CREF) which is subscribed to by most American universities for their faculties. These pension funds could be managed by the government, by private firms, or as joint ventures. The only requirement is that the funds invest in a balanced portfolio of government and private securities that yield a respectable rate of return and are kept insulated from irrelevant political pressures. As in TIAA/CREF, individuals may be permitted modest latitude in choosing among investment opportunities.

The point of the example is that prospective real resources are adequate to finance early retirement, expanded high quality education, and an increasing level of high quality medical care (I assume that the U.S. medical expenditures will increase to about 20 percent of Gross Domestic Product by 2040). The typical working household will still have 68 percent of a substantially larger income than is typical today to spend on other forms of consumption. Since current levels of food, clothing and shelter will require a decreasing number of hours of work during the family's life cycle, dropping to about 300 hours of work annually for the typical household by 2040, families nearing retirement will have several options. They will be able to increase their rate of accumulation in consumer durables and housing, or increase expenditure on such consumables as travel, entertainment and education, or further reduce hours of work, or retire before 55.

Embedded in my simulation is a suggestion for modernizing current government systems of taxation and expenditure. Close to half of what are called taxes are actually deferred income or forced savings. In these cases the government does not collect money for its own benefit but merely acts as an intermediary in order to insure that money needed for later use (such as retirement by individuals) is set aside for the stated purpose and then delivered to households when needed. The particular form of intermediation exercised by the US government, however, is quite peculiar. Instead of setting up an account in the name of the individual doing the saving, the government transfers the funds to a person who had earlier deferred consumption. At the same time it promises the current taxpayer that when he or she is ready to retire, the government will find new taxpayers to provide the promised funds. Under normal circumstances OECD governments provide this form of intermediation quite efficiently. The costs of administering the U.S. Social Security system, for example, is less than 1 percent of expenditures (U.S. Social Security Administration 1997).

The problem with the current system, aside from the fact that it gives the impression that personal savings are actually taxes, is that its operation is subject to heavy political buffeting. As a consequence, rates of return on the savings for deferred income are highly variable and often far lower than they would have been had they been invested in a fund similar to TIAA/CREF. Moreover, the current system is affected by variations in the fertility and mortality rates that have created financial crises and thrown into doubt the government's promises that it will be able to provide the money supposedly set aside for later retirement income, health care, or education.

The crisis then is not in a nation's resources for providing extended retirement, improved health care, and extended education, but in the exceedingly clumsy system for financing these services. The crisis is to a large extent due to accidents of history. When the original social security systems were established prior to World War I they were intended to be class transfers. The levels of transfers were modest, supplying the elderly with barely enough food to keep them from starving. Such payments were not generally expected to cover the cost of housing or other necessities of life. Moreover, only a small percentage of a cohort was expected to live long enough to become eligible for the benefits and the average duration of support was expected to last only a few years. Under these circumstances a tax of a few percent on the income of the rich was adequate to fund the program. The rich of Prussia and Great Britain were prepared to bear this cost for the sake of political stability (Fogel 1999, ch. 5).

Over the course of the twentieth century, however, the enormous increase in life expectancy and the rising standard of living led to much longer periods of retirement and much higher levels of support after retirement. Such programs could no longer be financed through a highly concentrated class tax. To support more expensive pension systems, taxes had to be extended to the entire working population. In so doing social security programs were transformed from redistribution schemes into systems of forced savings, although the transformation in the basic nature of these systems was obscure to most participants. Modernization of the essentially self-financed programs for retirement, health care and education from their current unsustainable systems of financing to a more transparent system of forced savings in provident funds is not easy, but it can be done in a manner that preserves intergenerational equity.

I have focused this analysis on the typical (median or average income) household in order to demonstrate that the economics of OECD nations have the prospective resources to permit early retirement, expanded education and expanded medical care. Unfortunately, the income of some households is so low that saving 32 percent of earnings would not provide a provident fund large enough to permit decent retirement, health care and education for these households. This is not a problem of inadequate national resources but of inequity. Such inequities can continue to be addressed by redistributing income from high-income to poor households by taxes and subsidies. Correcting these inequities does not require restricting retirement or health care.

THE DEMOCRATIZATION OF SELF-REALIZATION

Self-realization requires good health and extensive leisure. The process of technophysio evolution is satisfying these conditions. Self-realization also requires, however, an answer to the question that persons with leisure have contemplated for more than 2000 years. How do individuals realize their fullest potential? Technophysio evolution is making it possible to extend this quest from a minute fraction of the population to almost the whole of it. Although those who are retired will have more time to pursue this issue, even those still in the labor force will have sufficient leisure to seek self-realization either within their professional occupation, or outside of them (Laslett 1991; Lenk 1994).

One implication of this analysis is that decision-makers both in government and in the private sector now need to review existing policies for their bearing on the timely growth of institutions that will satisfy an expanding demand for leisure. Some may consider it premature to speculate on the new forms of human activity that will come into being simply to provide a better understanding of ourselves and our world. What is required is more than an expansion of existing universities and other forms of adult education. Entirely new educational forms are needed that aim at satisfying not only curiosity, but also a longing for spiritual insights that enhance the meaning of life, and that combine entertainment with edification and sociality. I believe that the desire to understand ourselves and our environment is one of the fundamental driving forces of humanity, on a par with the most basic material needs. We are lucky to be living in an age that provides vast amounts of time, much longer lives and better health to satisfy this urge. But we are also faced with a new set of equity issues—spiritual equity—that could become more divisive than those that threatened stability a century ago.

ACKNOWLEDGMENTS

Research on this paper was supported by NIH Grant AG10120, NSF Grant SES-9114981, and the Charles R. Walgreen Foundation.

REFERENCES

1. Allen, R. 1992. Enclosure and the yeoman: The agricultural development of the South Midlands, 1450-1850. Oxford: Oxford University Press.

2. Allen, R. 1994. Agriculture during the industrial revolution. In The economic history of Britain since 1700 (eds. R. Floud and D. McCloskey), 2nd ed., vol. 1, pp. 96-122. Cambridge: Cambridge University Press.

3. Ausubel, J.H. & Grübler, A. 1995. Working less and living longer: Long-term trends in working time and time budgets. Technological Forecasting and Social Change 50: 113-131.

4. Barker, D.J.P., ed. 1992. Fetal and infant origins of adult disease. London: British Medical Journal.

5. Barker, D.J.P. 1994. Mothers, babies and disease in later life. London: BMJ Publishing Group.

6. Bishop, C.W. 1936. Origin and early diffusion of the traction plow. Antiquities 10: 261-81.

7. Chavez, A., Martinez, C. & Soberanes, B. 1995. The effect of malnutrition on human development: A 24-year study of well-nourished children living in a poor Mexican village. In Community based longitudinal studies of the impact of early malnutrition on child health and development: Classical examples from Guatemala, Haiti and Mexico (ed. N.S. Scrimshaw), pp. 79-124. Boston: International Nutritional Foundation for Developing Countries.

8. Cipolla, C.M. 1974. The economic history of world population. 6th ed. Harmondsworth, Middlesex: Penguin.

9. Clark, J.G.D. 1971. World prehistory: An outline. Cambridge: Cambridge University Press.

10. Costa, D.L. 1998. The evolution of retirement: An American economic history, 1880-1990. Chicago: University of Chicago Press.

11. Czeizel, A.E. & Dudás, I. 1992. Prevention of the first occurrence of neural-tube defects by periconceptional vitamin supplementation. New England Journal of Medicine 327: 1832-1835.

12. Derry, T.K. & T.I. William. 1960. A short history of technology. London: Oxford University Press.

13. Fagan, B.M. 1977. People of the earth. 2nd ed. Boston: Little, Brown & Co.

14. Fogel, R.W. 1994. Economic growth, population theory, and physiology: The bearing of long-term processes on the making of economic policy. American Economic Review 84: 369-395.

15. Fogel, R.W. 1999. The Fourth Great Awakening: The political realignment of the 1990s and the future of egalitarianism. Chicago: University of Chicago Press (forthcoming).

16. Fogel, R.W. & Costa, D.L. 1997. A theory of technophysio evolution, with some implications for forecasting population, health care costs, and pension costs. Demography: 34, 49-66.

17. Fogel, R.W. & Floud, R. 1999. A theory of multiple equilibria between populations and food supplies: Nutrition, mortality and economic growth in France, Britain and the United States, 1700-1980. Typescript, University of Chicago.

18. Fogel, R.W., Costa, D.L. & Kim, J.M. 1993. Secular trends in the distribution of chronic conditions and disabilities at young adult and late ages, 1860-1988: Some preliminary findings. Paper presented at the NBER Summer Institute, Economics of Aging Program, 26-28 July, Cambridge, Mass.

19. Iyer, S.N. 1993. Pension reform in developing countries. International Labour Review 132: 187-207.

20. Laslett, P. 1991. A fresh map of life. Cambridge: Harvard University Press.

21. Lee, C. 1996. Essays on retirement and wealth accumulation in the United States, 1850-1990. Ph.D. dissertation, University of Chicago.

22. Lenk, H. 1994. Value changes and the achieving society: A social-philosophical perspective. In OECD societies in transition: The future of work and leisure, pp. 81-94. Paris: OECD.

23. Lozoff, B., Jimenez, E. & Wolf, A.W. 1991. Long-term developmental outcome of infants with iron deficiency. New England Journal of Medicine. 325:687-695.

24. Maddison, A. 1991. Dynamic forces in capitalist development. Oxford: Oxford University Press.

25. Manton, K.G. 1993. Biomedical research and changing concepts of disease and aging: Implications for long-term forecasts for elderly populations. In Forecasting the health of elderly populations (ed. K.G. Manton, B.H. Singer & R.M. Suzman), pp. 319-365. New York: Springer-Verlag.

26. Manton, K.G., Corder, L. & Stallard, E. 1997. Chronic disability trends in elderly United States populations: 1982-1994. Proceedings of the National Academy of Sciences, USA 96: 2593-2598.

27. Martorell, R., Rivera, J. & Kaplowitz, H. 1990. Consequences of stunting in early childhood for adult body size in rural Guatemala. Annales Nestlé 48: 85-92.

28. McNeill, W. 1971. A world history. 2nd ed. New York: Oxford University Press.

29. Piggot, S. 1965. Ancient Europe from the beginnings of agriculture to classical antiquity. Chicago: Aldine.

30. Poortvliet, W.G. & Laine, T.P. 1995. A global trend: Privatization and reform of social security pension plans. Benefits Quarterly 11 (3): 63-84.

31. Rosenberg, I.H. 1992. Folic acid and neural-tube defectsCtime for action? New England Journal of Medicine 327: 1875-1877.

32. Scrimshaw, N.S. 1993. Malnutrition, brain development, learning and behavior. The Twentieth Kamla Puri Sabharwal Memorial Lecture presented at Lady Irwin College, New Delhi, 23 November.

33. Scrimshaw, N.S. & Gordon, J.S., eds. 1968. Malnutrition, learning and behavior. Cambridge: MIT Press.

34. Slicher von Bath, B.H. 1963. The agrarian history of Western Europe A.D. 500-1850. London: Edward Arnold.

35. Tanner, J.M. 1990. Foetus into man: Physical growth from conception to maturity. Rev. ed. Cambridge: Harvard University Press.

36. Tanner, J.M. 1993. Review of D.J.P. Barker's Fetal and infant origins of adult disease. Annals of Human Biology 20: 508-509.

37. Trewartha, G.T. 1969. A geography of populations: World patterns. New York: John Wiley & Sons.

38. Wrigley, E.A. 1987. Urban growth and agricultural change: England and the continent in the early modern period. In People, cities and wealth: The transformation of traditional society, pp. 157-193. Oxford: Basil Blackwell.

2. Prevention of Aging

THE BIOLOGY OF AGING AND LONGEVITY

Calvin B. Harley

Geron Corporation
200 Constitution Drive
Menlo Park, CA 94025
USA

INTRODUCTION

Issues of aging and longevity are intertwined in all aspects of human life. From a biological perspective, the search for the underlying mechanisms is a fascinating problem for which there is little consensus regarding the best approach. Although research in this area is as old as science itself, meaningful insights into the fundamental causes of aging and longevity at the molecular or genetic levels are very recent and come from dramatically varied model systems. Thus, it is not surprising that one will find almost as many theories of aging as there are scientists working on this problem.

"Aging" is commonly thought of as the characteristic set of changes which occur with time after development to adulthood. The deleterious aspects of aging contributing to morbidity and the increasing probability of death can be referred to as "senescence". However, aging and senescence are generally used interchangeably, and at the molecular level there may be little reason to make a distinction. The relationship between longevity, or maximum lifespan, and aging is not a simple one. However, it is clear that genetics play a major role in determining the species-specific characteristics of aging and longevity. Even within a species, there are strong genetic effects on aging and longevity which are more significant than most environmental effects.

This review highlights some of the significant new discoveries in the field of aging and longevity research, with a particular focus on what we are learning about the genes and cellular events that contribute to aging in humans. In particular,

Longevity and Quality of Life, Edited by Butler and Jasmin
Kluwer Academic / Plenum Publishers, New York, 2000.

about the genes and cellular events that contribute to aging in humans. In particular, investigations into the difference between immortal cells (from cancer and germline tissues) and their mortal somatic counterparts have revealed a fundamental understanding of aging in human cells and opened a variety of novel approaches for the diagnosis and treatment of age-related conditions, including cancer.

The hallmarks of mortality in dividing human somatic cells are a finite division capacity (the "Hayflick Limit") and alterations in gene expression causing loss of tissue function (1,2). The telomere hypothesis of cell aging (3) can explain the fundamental characteristics of cell mortality and immortality. Telomeres are essential genetic elements at the ends of our chromosomes which are maintained by a specialized enzyme called telomerase (4). We have recently proven that introduction of the gene for the catalytic sub-unit of telomerase into normal cells can halt the cellular aging process in the laboratory and confer a long and healthy life span to these cells (5). This and related biological research have the potential for significant impact on human health and longevity.

DEFINITIONS.

Aging can be defined as progressive, universal, and normally irreversible changes which occur over time post-maturation. The genetic basis for these changes are set at conception, and at the molecular level they may begin long before clearly adaptive developmental changes are over. However, it is difficult to know whether events that are normally associated with aging late in life may actually be part of the adaptive program of growth and development early in life. For example, collagen cross-linking and mineral deposition can strengthen connective tissues as we mature, but when they continue late into life, these processes can create stiffening of joints and reduced diffusion of oxygen and metabolites throughout the body, thus contributing to pathological changes. Similarly, a genetic program which reduces the probability of cancer by conferring a finite lifespan to most normal body cells is advantageous early in life but deleterious late in life. In the first few decades of life, the finite replicative capacity of cells would not be reached in normal tissues, but would be exhausted in many developing cancers, thus limiting their growth. However, late in life when some normal cells reach the end of their genetically programmed lifespan, this advantageous early trait contributes to age-related decline in organ and tissue function. Thus, to distinguish aging from development, it is convenient to define the onset of age-related changes at maturation.

Some signs of age-related changes, such as graying of hair and the wrinkling of skin, are typically not associated with pathology, even though many will suffer emotionally from their onset. Other changes, such as declining testosterone levels in males, may be welcome events by some, as otherwise aggressive personalities mellow out. The causes of such "neutral" or even "positive" aging changes are likely to be closely related to those of life-threatening changes elsewhere in the body, and it may not be useful to attempt to distinguish "aging" from "senescence" at the molecular or genetic level.

Can we distinguish aging from disease? Clearly, certain infectious diseases and genetic disorders are not associated with aging per se, aging can occur for decades without disease, and no overt, clinically diagnosed disease is truly universal. Thus, aging is distinct from disease. Nevertheless, it is useful to refer to age-related diseases as those whose incidence increases with age, and which are believed to be the clinical manifestation of the underlying senescence occurring in our cells, tissues and organs. Thus, atherosclerosis, osteoporosis, cancer, and even pneumonia in the elderly can be considered age-related in the sense that declining functions in aging endothelial cells (atherosclerosis), osteoblasts (osteoporosis), and immune, mucosal and connective tissue cells (cancer and pneumonia) are causally involved in onset of these diseases. Of course, which disease or diseases any one individual will succumb to is difficult to predict with today's technology; it will depend upon the individual's exact genetic makeup and set of experiences. However; age-related diseases can usually be identified in any very old human.

Longevity is commonly thought of as the maximum lifespan ever observed for an individual in a species. However, recent work has shown that this parameter is highly sensitive to the size of the population investigated, and that mortality rates in the oldest old are significantly lower than those in the rest of the population (see J. Vaupel). This begs the question of whether "longevity" can even be defined for a species. Nevertheless, an arbitrary definition of longevity based on, for example, the age by which 99.9 percent of the individuals in a large population have died, is a useful tool in characterizing the greatest lifespan most members in a species could hope to achieve with optimal health care, short of significant gene therapy.

These definitions are simply a matter of convenience for the purpose of discussion. Biologists need to focus on understanding the underlying molecular causes of the changes which contribute to increased morbidity and mortality with time. It is anticipated that with increased funding for research and development on the biology of aging we will have the opportunity to develop new medicines to significantly increase our "healthspan," if not our longevity.

THEORIES: GENETIC DETERMINANTS AND THE ENVIRONMENT

It has often been stated that there are two major camps of biological research on aging: the "nature" camp, where one studies genes that influence aging, and the "nurture" camp, where the focus is on physical, or entropic changes such as oxidation, glycation, or mutation. However, genetic and environmental factors are so intimately related that it makes little sense to think of them separately. Many genes known to influence longevity or disease are involved in preventing, repairing, or modulating the rate of physical damage to macromolecules arising from various internal and external factors. The interplay between genes and the environment in fact can give the impression of a genetic program for aging. However, from an evolutionary perspective, it is highly unlikely that aging is programmed or adaptive, even though it is genetically determined. Many evolutionary biologists think of aging as a complex set of processes which were genetically programmed for their adaptive, positive effects early in life, but which happen to have deleterious unselected effects late in life. The finite lifespan of normal somatic cells could be one such process, where the genetically programmed advantage early in life is a reduced probability of cancer, but the deleterious consequence late in life is an increasing fraction of old non-dividing cells with detrimental behavior at sites of chronic stress.

As we gain greater knowledge of the fundamental processes in human aging, it can be expected that the overlap between genetic and environmental factors will become clearer. For example, damage from free radicals or mutations resulting in cell death impacts on replicative aging by increasing cell turnover and causing an accelerated exhaustion of the replicative limit as a function of time. An even more direct link between oxidative damage and the "mitotic clock" was demonstrated with the finding that increased oxygen stress causes cells to lose more telomeric DNA with each cell division, dramatically hastening the onset of replicative senescence (6). However, even with this synthesis of aging theories, there are still a multitude of sub-theories, arising in part from the diversity of model systems under investigation.

EXPERIMENTAL MODELS OF LONGEVITY

Numerous models of aging have been studied utilizing convenient laboratory organisms such as yeast, fruit flies, round worms, and rodents. The relevance to humans of age-related changes and genes which influence longevity in these organisms remains to be determined. There are some intriguing similarities between species in proposed mechanisms of aging, but the differences in lifespan, life history, the nature of pathological change, and the distribution of dividing and non-dividing cells make it extremely difficult to extrapolate from one species to another. It can also be argued that the relative lack of evolutionary pressure on genetic effects late in life will lead to significant divergence in aging mechanisms between species. Even comparing mice to humans has its limitations, where, for example, the difference in telomere length between these species leads to dramatically different time scales for the onset of telomere-related defects (7). In primates, understandably little experimental research has been conducted to evaluate potential genetic or environmental factors influencing longevity. Long-term studies of caloric restriction are underway in monkeys (8, 9). Caloric restriction is currently the only known manipulation which significantly and reproducibly decreases the rate of aging in a mammalian species.

In humans, the search to identify genes influencing aging and age-related disease have taken several tacks. The study of a rare accelerated aging condition called Werner's Syndrome led to the identification of the gene responsible for this genetic defect (10). However, there is no evidence yet that this gene plays a role in normal aging. Similarly, searching candidate genes for polymorphisms (genetic variants) commonly found in centenarians led to the identification of certain variants of the genes for apolipoprotein E and the angiotensin converting enzyme as longevity-associated (11). However, further research is again needed to understand the significance and mechanisms of putative "longevity-assurance" genes. A third approach utilizing human cellular systems has recently proven productive in identifying a key gene involved in cell mortality and immortality with implications in aging and cancer.

CELL MORTALITY

The finite lifespan of normal human body cells (1) is now understood in terms of genetic events at the telomeric tips of our chromosomes (3, 4). Telomeres, which are comprised of a specific repeated DNA sequence (TTAGGG in humans) and associated proteins, have been compared to the plastic tips at the end of a shoelace; they protect the chromosome end from damage. However, telomeric DNA requires the action of a special enzyme, telomerase, to maintain their full length when the cell divides. Telomerase synthesizes the TTAGGG repeats of the human telomere (12). In the absence of telomerase, telomeres gradually shorten with replicative age, both in the laboratory, and in the body (13, 14, and references therein). A large body of literature points to the absence or very low levels of telomerase in mortal somatic cells which senesce at a characteristic threshold telomere length, and the readily detected levels of telomerase in immortal cells found within the reproductive germline and in cancers of all types. Cells with telomerase generally maintain a constant telomere length which can be long (typical of germline cells) or short (typical of cancer cells).

CELL LIFESPAN EXTENSION

To confirm that the limited lifespan of normal human cells was in fact dictated by shortening telomeres in the absence of telomerase, we introduced the gene for the telomerase catalytic subunit hTERT under a strong promoter (start signal) into normal human cells which have the endogenous hTERT gene repressed. We had earlier demonstrated that an active copy of the hTERT gene was capable of reconstituting telomerase activity in normal telomerase-negative cells. The active telomerase gene halted or reversed telomere loss in the transfected cells and has increased their lifespan, at the time of this publication, by at least 4-fold. In addition, transfected cells maintain a youthful phenotype and have shown no signs of chromosome instability or malignant transformation. These data indicate that telomerase is capable of preventing cellular senescence without causing loss of growth control or differentiated function. Thus, telomerase is clearly distinct from typical cancer genes. Moreover, it is clear that unless telomerase has additional properties related to the prevention or correction of generalized macromolecular damage, the only limiting factor in normal replicative aging is telomere loss.

CLINICAL IMPLICATIONS AND FUTURE DIRECTIONS

Replicative senescence of normal somatic cells limits their functional capacity both inside the body and *ex vivo*, for example when they are removed for purposes of cell therapy. Conversely, escape from replicative senescence, i.e., cellular immortality, is a hallmark of tumor cells, especially those that progress to malignancy. Thus, the obvious sites of intervention are regulated telomerase activation or gene therapy to increase the youthful lifespan of aging cells, and the inhibition of telomerase to "remortalize" tumor cells. Given the highly specific association of telomerase activity with tumors, there is also a number of diagnostic applications for telomerase assays. Until more is known about the consequences of therapeutically regulating telomerase activity in humans, research and development in the next few years will largely focus on cellular and animal studies. However, it is expected that therapeutic and diagnostic products based on telomerase will be tested in human clinical trials for the treatment of disease in the foreseeable future.

Clinically relevant cell types which might first be studied *ex vivo* for transplant use include hematopoietic stem cells (following high-dose chemotherapy), lymphoid cells (AIDS), skin cells (grafting), retinal pigmented epithelial cells (age-related macular degeneration), and endothelial cells (restenosis). All of these cell types are known to possess a finite replicative capacity and lose telomeric DNA with age. *In vivo* application of telomerase activation will necessarily involve greater testing, but could be applied to the same cell types and others, such as chondrocytes (osteoarthritis) and osteoblasts (osteoporosis). In addition, telomerase-immortalized normal human cells are expected to be useful in a number of applications in the biopharmaceutical industry, including drug discovery and therapeutic protein production. Even as these opportunities are being explored, research is ongoing to more fully characterize the link between replicative aging *ex vivo* and disease pathology *in vivo*.

The years ahead promise to be exciting for all involved in applying our new insights into the mechanisms of human aging to the development of new medicines to address disease and longevity.

REFERENCES

1. Hayflick, L., and Moorhead, P. S. (1961). The serial cultivation of human diploid cell strains. Exp. Cell Res. 25, 585-621.

2. West, M. D. (1994). The cellular and molecular biology of skin aging. Arch. Dermatol. 130, 87-95.

3. Harley, C. B. (1991). Telomere loss: mitotic clock or genetic time bomb? Mut. Res. 256, 271-282.

4. Blackburn, E. H. (1991). Structure and function of telomeres. Nature 350, 569-573.

5. Bodnar, A. G., Ouellette, M., Frolkis, M., Holt, S. E., Chiu, C.-P., Morin, G. B., Harley, C. B., Shay, J. W., Lichtsteiner, S., and Wright, W. E. (1998). Extension of life-span by introduction of telomerase into normal human cells. Science 279, 349-352.

6. von Zglinicki, T., Saretzki, G., Docke, W., and Lotze, C. (1995). Mild hyperoxia shortens telomeres and inhibits proliferation of fibroblasts: a model for senescence? Exp. Cell Res. 220, 186-193.

7. Lee, H.-W., Blasco, M. A., Gottlieb, G., Horner II, J. W., Greider, C. W., and DePinho, R. A. (1998). Essential role of mouse telomerase in highly proliferative organs. Nature 392, 569-574.

8. Roth, G. S., Kowatch, M. A., Hengemihle, J., Ingram, D. K., Spangler, E. L., Johnson, L. K., and Lane, M. A. (1997). Effect of age and caloric restriction on cutaneous wound closure in rats and monkeys. J. Gerontol. 52A, B98-B102.

9. Ingram, D. K., Cutler, R. G., Weindruch, R., Renquist, D. M., Knapka, J. J., April, M., Belcher, C. T., Clark, M. A., Hatcherson, C. D., and Marriott, B. M. (1990). Dietary restriction and aging: the initiation of a primate study. J. Gerontol. 45, B148-63.

10. Yu, C.-E., Oshima, J., Fu, Y.-H., Wijsman, E. M., Hisama, F., Nakura, J., Miki, T., Ouasis, S., Martin, G. M., Mulligan, J., and Schellenbergt, G. D. (1996). Positional cloning of the Werner's syndrome gene. Science 272, 258-262.

11. Schachter, F., Faure-Delanef, L., Guenot, F., Rouger, H., Froguel, P., Lesueur-Ginot, L., and Cohen, D. (1994). Genetic associations with human longevity at the APOE and ACE loci. Nature Genet. 6, 29-32.

12. Morin, G. B. (1989). The human telomere terminal transferase enzyme is a ribonucleoprotein that synthesizes TTAGGG repeats. Cell 59, 521-529.

13. Harley, C. B., Futcher, A. B., and Greider, C. W. (1990). Telomeres shorten during ageing of human fibroblasts. Nature 345, 458-460.

14. Chang, E., and Harley, C. B. (1995). Telomere length and replicative aging in human vascular tissues. Proc. Natl. Acad. Sci. USA 92, 11190-11194.

15. Shay, J. W., and Bacchetti, S. (1997). A survey of telomerase activity in human cancer. Eur J Cancer 33, 777-791.

16. Shay, J. W., and Wright, W. E. (1998). Telomeres and telomerase in the regulation of human cellular aging. In Alfred Benzon Sumpsoium 44 (Copenhagen, Demark: MUNKSGAARD).

QUALITY OF LIFE AND DISORDERS OF LONGEVITY

Francoise Forette

Hopital Broca
54-56 rue Pascal
Paris 75013
France

INTRODUCTION

Like all western countries, France is facing an extraordinary longevity revolution. The mean life expectancy is now 73 years for men and 82 years for women. Around 20 percent of the population is over sixty years of age. In 2015, 40 percent of the population will be over fifty. Nowadays, only one third of the population over fifty is still working. Projections suggest that in 2015, 30 percent of the population will have to financially support 70 percent of a not yet professionally active younger population and of retired or unemployed older persons.

The ethical question we have to address is: How can we maintain equity and promote the quality of life of all generations living together?

The main determinants of quality of life over the age of 60 are satisfactory health status, financial autonomy, family links, positive image, social role and personal responsibility. The ethical challenge is to ensure that economic pressure does not compromise the rights of this expanding population to highly skilled medical care (particularly in preventive areas), to fair retirement pensions or access to work, to political influence, to a social role including access to culture, and to a productive life in aging.

All medical data demonstrate that a vast majority of the aging population is going to enjoy remarkable conditions of health, autonomy and productivity. This rising tide of candidates for a burgeoning longevity will rapidly expand with the baby-boomer generation. This generation, being the largest, has deeply influenced our societies and will certainly change the image of age and aging.

On the other hand, two critical issues must be considered. First, a minority of elderly people remain at risk of contracting debilitating chronic diseases that lead to a devastating dependence, but research and prevention should progressively decrease these occurrences. Ethics and morality require a particular vigilance toward this small minority, who are victims of chronic diseases often associated

Longevity and Quality of Life, Edited by Butler and Jasmin
Kluwer Academic / Plenum Publishers, New York, 2000.

with unfavorable sociological conditions which may lead to a "no care zone" (See Estes 1993 for a review).

The second critical issue is the existence of a growing number of persons who happily reach a very advanced age—ninety, ninety-five even a hundred years. These nonagerians and centenarians, though often healthy, remain a frail population, and require assistance to remain autonomous.

Innovative answers must be found for these two phenomena but they are marginal when compared to the increasing number of dynamic, healthy and productive aging individuals.

QUALITY OF LIFE RELATED TO HEALTH STATUS AND HIGHLY SKILLED MEDICAL CARE

The compression of morbidity into the last few years of life hypothesized by Fries (1983) is no longer out of reach. The disability-free life expectancy is increasing more rapidly than the mean life expectancy, as shown by Robine and Ritchie.

The self-reported prevalent morbidity is still high in elderly persons. In a recent French study on a representative sample of people 65 and over, the mean number of self-reported diseases is 7.6 per person. The number is more important in women, 8.4 per person, than in men, 6.8 per person. (Sermet 1994).

But a declining incidence of most conditions leading to disability and a decrease in institutionalization is observed in Europe as well as in the US (Manton 1995). Stroke represents a very good example of a debilitating disease, the decline of which is multifactorial (Whisnant 1984). However, it is clear that a better control of the identified risk factors will strengthen this trend. Community control has been highly effective in Europe in the North Karelia Project (Vartiainen 1994).

Prevention opens up avenues of progress in defeating many conditions. Besides cardiovascular diseases, osteoporosis and fractures are a leading cause of death, disability and institutionalization. In women, one of the most effective preventive measures is estrogen replacement after menopause. In France, the total percentage of treated women is less than 10 percent. But if you look at women who have recently become menopausal, the percentage rises to 30 percent. Although progress has been made, enormous advances remain to be made. As underscored by Manton, "disease prevention, a delay in age at onset, or a disease cure all can cause morbidity prevalence to decline... For example, slowing the development of cataracts (e.g., by use of antioxidants) and increasing the age at which they appear by 10 years, would reduce their prevalence by half..." (Taylor 1993).

Dementias of all causes still represent the most striking factor of impairment of quality of life after 60. There again, progress in research may radically change the painful fate of many elderly patients.

The last point concerns the links between quality of life, health status and highly skilled medical care. It is well known that health status depends on various factors: genetic and biological, socioeconomic level, education, social security systems, etc. But it has also been proven that access to care and quality of care are major determinants for improving the health status of older people and particularly, the frail elderly. Specific geriatric programs have been evaluated in the US and a number of countries (Stuck 1933; Stuck 1995). Most of these studies have confirmed that Comprehensive Geriatric Assessment (Rubenstein 1988; Rubenstein 1995), Geriatric Evaluation and Management Units, and departments or divisions of geriatric medicine not only decrease early mortality but also improve the functional status of the frail elderly, reduce the frequency of discharge to institutions for long-term care, and decrease the rate of hospital readmission. This is probably the best way to promote quality of life.

While nearly all countries have to face increasing pressures for cost containment, it is interesting to point out that specific geriatric care does not inflate care cost and is proven cost-effective over the long term (Rubenstein 1995). It may even induce a significant reduction in the cost of hospitalization (Naughton 1994). It is clear this is the key to better health for our aging population, and therefore a key to a better quality of survival for the frail elderly.

QUALITY OF LIFE RELATED TO ECONOMIC ISSUES

Access to work or fair retirement pensions are another key to quality of life. France made the decision to lower the retirement age in 1981 for two reasons: First, it was an old claim laid, for many years, by the unions and it was written in the Socialist program for the 1981 presidential election. Secondly, politicians thought that this measure could be a miraculous cure for unemployment. However, France has been a model to all western nations in clearly showing that this is not the case. Indeed, early retirement is appropriate for a very small fraction of the population whose careers are the longest, the toughest, and the most exposed to pathogenic factors. But when applied to the whole population, it certainly is the most demagogic, unconsidered and irresponsible decision ever reached by politicians.

The present economic crisis has hit an increasing number of aging workers who have progressively been rejected from production or confined to marginal or virtually obsolete positions. Retirement is therefore imposed earlier on people who do not always want it and who are rarely prepared for it. This devaluation fundamentally contradicts the image of an active new beginning in life when one ages.

The incomes of older persons have nevertheless greatly increased in France for the past 20 years. In the seventies, they were 20 percent lower than that of the younger populations. In 1996, the situation was reversed, and today the financial resources of the elderly population are at least 5 percent higher. That is the case only for the “young” elderly up to 75 years. Around 10 percent of the population over 65 live on social welfare. Most of them are widowed, single or divorced and over 75 years. But, as a whole, the elders are wealthy (David 1994).

Economic studies always focus their interest on the “cost” of the aging process. Now, with the emergence of a large dynamic, healthy and wealthy elderly population we have to look at the other side of the same question. We must determine the contribution of the aging population to the economy of our countries. Money is often transferred from the eldest to the youngest. Grandparents help their children to buy houses, their grandchildren to get an education, etc. (Attias Donfut 1996; Poquet 1996).

Older persons participate more than ever in the economic cycle, due to their increase in numbers and purchasing power. Housing expenses are equivalent to those of other age categories; food expenditures are slightly higher, and health expenditures are significantly higher. In the same vein, clothing, education, and leisure expenses which were classically lower now tend to increase and represent an important part of the budget.

As a whole, it is clear that aging people still represent an important part of the economic system and probably a specific "market."

Even the frail elderly contribute to the economic circuits. When they are dependant and need assistance, they generate new jobs in European countries where unemployment is a devastating problem.

If we succeed in emphasizing the economic role of older persons it will radically change the image of aging. Studies such as Charlotte Muller's (1994, 1995) looking at the influence of retirement age on the well being, feeling of accomplishment and health conditions of retired persons tend toward a "reconciliation of individual preferences and potentials of the current job structure."

QUALITY OF LIFE RELATED TO SOCIAL ROLE AND FAMILY

Transfers from generation to generation are not limited to monetary transfers and the ILC study analyzing "Legacy and Responsibility Among Generations" also underlines what philosophical or moral values the generations want to pass on, such as family ties, compassion, religious spirit, freedom, democracy, etc. Then, the youngest generations may take stock of the fantastic legacy left by their forefathers. In France, a project run by the National Foundation of Gerontology encourages schoolchildren to gain an awareness of the attainments which are the privileges of age. The title of the project is "Growing up is Aging, Aging is Growing up" (Arfeux-Vaucher 1996).

New data from the Swedish study on "Seventy-year-old people in Gothenburg" address the issue of "Life events and quality of life in old age" (Grimby 1995). Among the negative life events which may strongly impair quality of life, children's divorcing parents is ranked as the most severe, the loss of spouse as the second most severe and personal illness is third. Family quarrels and conflicts, and breaking off a relationship were also associated with poor quality of life. Reports of positive life events included birth of a grandchild, recovery from illness, recovery of a spouse or finding a new partner. It is clear from this study that family links and personal or spousal health status are major determinants of well-being.

QUALITY OF LIFE AND ATTITUDES TOWARD AGING

In *The Fountain of Age*, Betty Friedan (1993) describes how our own fear of aging leads to a catastrophic image of this normal process, viewing it as a "problem" instead of an accomplishment. She underlines that "there seemed to be a growing impatience for some final solution to that problem before the multiplying numbers of invisible, unproductive, dependant older people, unfortunately living beyond sixty-five, placed an intolerable burden on their families and societies with their senility, chronic illnesses, Medicare, Meals on Wheels, and nursing homes." She also emphasizes that we gerontologists contribute to this negative image because we

Antonini (1988) remarkably analyzed that the creative capacities of the painter Monet were magnified and not impaired by the emergence with age of a physical deficit. The chronological sequence of Monet's Nymphéas shows that the loss of structure of the forms as well as the monochrome tendency appears concurrently with an increase of visual deficit. These paintings constitute striking proof that physical depredation did not lead to a creative decline, but rather, determined a change of style which placed Monet among the precursors of nonfigurative art.

It is clear that the satisfactory health status of the aging population will contribute to promoting the image of productive aging. But productive aging is possible even in handicapped, frail elderly. Indeed, they invent fantastic compensatory strategies, judiciously use their remaining capacities, circumvent problematic situations, and transform handicap into gain. This is how the aging build up their own quality of life.

WHAT IS QUALITY OF LIFE AND DO WE KNOW HOW TO MEASURE IT?

This is a question that ought to have been addressed first. Slow but important progress in assessment of quality of life is being made (Testa 1996). Tools are being developed, mainly in the health field (Grimby 1995; Fergus 1996). A number of these tools are specifically focused on the geriatric population (Guyatt 1993; Rai 1995). We are able to analyze and measure some of the physical, psychological and social determinants. We still cannot define quality of life because its origin is buried in the essence and mystery of human nature.

REFERENCES

1. Antonini, FM & Magnolfi, S (1988). Créativité et vieillissement. Gerontologie 68, 23-34.

2. Arfeux-Vaucher, G (1996). Grandir et Vieillir : Parlons ans. Paris, France: Fondation Nationale de Gérontologie.

3. Attias-Donfut, C (1996). Les solidarités entre générations. In La société fançaise. Données sociales. Paris, France : INSEE.

4. Brymer C.D, Kohm C.A, Naglie G, Shekter-Wolfson L, Zorzitto M.L, O'Rourke K (1995). Do geriatric programs decrease long-term use of acute care beds? Journal of the American Geriatric Society 43, 885-889.

5. David M.G & Starzec, C (1994). Les revenus des personnes de 60 ans et plus : Structures et disparités. Gerontologie et société . Cahiers de la Fondation Nationale de Gérontologie, 71,181-197.

6. Estes CL, Swan, J &Associates (1993) The Long term care crisis. Elders trapped in the no-care zone. Newbury Park, London, New Dehli: Sage Publications.

7. Fergus I, Demopoulos L.A, Lejemptel T.H (1996). Quality of Life in Older Patients with Congestive Heart Failure. Drugs & Aging. 8,;23-28.

8. Friedan B, (1993). The Fountain of Age. New York: Simon & Schuster Publis.

9. Fries J.F. (1983) The compression of morbidity. Milbank Memorial Fund Quaterly/ Health and Society. 61, 397-419.

10. Grimby A, Svanborg A, (1996). Life events and the quality of life in old age. Report from a medical-social intervention study. Aging Clinical and Experimental. Research. 8 ,162-169.

11. Grimby A.U & Rosenhall U (1995). Health-related quality of life and dizziness in old age. Gerontology 41, 286-298.

12. Guyatt G.H, Eagle D.J, Sackett B, Willan A, Griffith L, McIlroy W, Patterson C.J, Turpie I, (1993). Measuring quality of life in the frail elderly. Journal of Clinical Epidemiology. 46,1433-1444.

13. Lipski P (1995). Optimum care of the elderly in an acute general hospital. Acute geriatric units hold the key to better health for our aging population. Medical Journal of Australia 164, 5-6.

14. Manton K, Sallard E & Corder L (1995). Changes in morbidity and chronic disability in the US Elderly Population : Evidence from the 1982,1984, and 1989 National Long Term Care Surveys. Journal of Gerontology : Social Sciences 50B, S194-S204.

15. Miller D.K, Lewis L.M, Nork & MJ, Morley JE (1996). Controlled trial of a geriatric case-finding and liaison service in an emergency department. Journal of American Geriatric Society 44, 513-520.

16. Muller, C. (1994). Seniors scientists and engineers. A study in productive aging. New York: International Longevity Center (US).

17. Muller, C & Sylver, C (1995). Generation project : A cross-cultural analysis of values and value transmission in the family : The case of Japan and the United States . Tokyo, Japan : International Longevity Center - Japan.

18. Naughton B, Moran M.B, Feinglass J, Falconer J & Williams, M(1994). Reducing hospital costs for the geriatric patient admitted from the Emergency Department: A Randomized Trial. Journal of the American Geriatric Society, 42 : 1045-1049.

19. Poquet G, Rochefort R (1996). Le pouvoir et le rôle économique des plus de 50 ans. Credoc. Paris, France : International Longevity Center -France.

20. Rai, G.S, Jetten E, Collas D, Hoefnagels W, Froeling P, Bor H (1995). Study to assess quality of life (morale and happiness) in two continuing care facilities- a comparative study in the UK and the Netherlands. Archives of Gerontology and Geriatrics. 20, 249-253.

21. Reuben D.B, Borok G.M, Wolde-Tsadik G, Ershoff D.H, Fischman L.K, Ambrosini V.L, Liu Y, Rubenstein L.Z, Beck J.C, (1995). A randomized controlled trial of comprehensive geriatric assessment in the care of hospitalized patients. New England Journal of Medicine. 332, 1345-1350.

22. Robine, J.M. & Ritchie K . (1991) Healthy life expectancy : evaluation of a global indicator of change in population health. British Medical Journal. 302, 457-60.

23. Rubenstein, L.Z., Wieland, D., Bernabei, R. (1995). Geriatric Assessment Technology Milano, Italy : Kurtis.

24. Rubenstein, L.Z., Wieland, D, Jopheson, KR, Rosbrook, B, Sayre J & Kane, RL (1988). Improved survival for frail elderly inpatients on a geriatric evaluation unit (GETU). Who benefits? Journal of Clinical Epidemiology. 41 : 441-449.

25. Rubenstein, L.Z.., Josephson, KR, Harker, JO, Miller, DK & Wieland D, (1995). The Sepulveda GEU Study revisited : Long-term outcomes, use of services and cost. Aging Clinical and Experimental. Research. 7 : 212-217.

26. Sermet C (1994) . La pathologie chez les personnes âgées en 1992. Gerontologie et société . Cahiers de la Fondation Nationale de Gérontologie, 71 : 24-41.

27. Stuck, A, Siu,A L, Wieland, D, Adam & Rubenstein, LZ. (1993) Comprehensive geriatric assessment : a meta-analysis of controlled trials. The Lancet 342 : 1032-1036.

28. Stuck A.E, Aronow H.U, Steiner A Alessi C.A, Büla C.J, Gold M.N, Yuhas K.E, Nisenbaum R, Rubenstein L.Z, Beck J.C, (1995). A trial of annual in home comprehensive geriatric assessments for elderly people living in the community. New England Journal Medicine 333: 1184-89.

29. Taylor A (1993). Cataract : Relationships between nutrition and oxydation. Journal of the American College of Nutrition. 12 : 138-146.

30. Testa, MA & Simonson, DC (1996). Assesment of quality of life outcomes. New England Journal of Medicine. 334 : 835-840.

31. Vartiainen, E, Puska, P, Pekkanen J, Tuomilheto, J & Jousilahti, P (1994). Changes in Risk factors explain changes in mortality from ischaemic heart diseases in Finland. British Medical Journal 309 : 123-7.

32. Whisnant, JP (1984) : The decline of strokes. Stroke 15 :160-168.

UNDERSTANDING THE BIOLOGICAL DETERMINANTS OF LONGEVITY: NEW OPPORTUNITIES AND CHALLENGES

Thomas Kirkwood

University of Manchester
3.239, Stopford Building
M139PT, Manchester
United Kingdom

INTRODUCTION

The major increases in human life expectancy that have occurred during the last two centuries (chiefly as a result of improvements in public health, housing, nutrition and general living standards) have taken place without any detailed knowledge of the biological determinants of longevity. These improvements have resulted in a general increase in survival during early life, through modification of extrinsic hazards which previously caused many premature deaths. The changes have been experienced most dramatically in developed countries but are also being experienced in the developing countries to a significant extent. Now that a greater fraction of the population is living to ages when longevity is limited by intrinsic biological constraints, it is becoming clear that further major changes in life expectancy are likely to depend on our understanding of the biological basis of aging.

NATURAL SELECTION AND AGING

The fact that human longevity in pre-industrial societies was strongly affected by extrinsic hazards, as opposed to the intrinsic frailty that comes with old age, conforms with a general pattern in the animal kingdom and tells us something very important about the nature of the aging process. In the wild, aged organisms are extremely rare because most animals die young from accidents such as being eaten by a predator, catching an infectious disease, or dying of starvation. Because old age is a rarity, any idea that the aging process has been programmed through natural

Longevity and Quality of Life, Edited by Butler and Jasmin
Kluwer Academic / Plenum Publishers, New York, 2000.

selection as part of an intrinsic death mechanism is likely to be false. Natural selection is relatively powerless to influence events that occur at older ages because older organisms are no longer around in sufficient numbers for selection to make itself felt.

The recognition of this principle, which was due chiefly to the work of J.B.S. Haldane, P.B. Medawar, and G.C. Williams, has provided important insights into the biology of aging. (Kirkwood & Rose 1991) Aging is thought to have evolved partly because natural selection cannot oppose the accumulation within the genome of late-acting deleterious mutations, and partly because of evolutionary trade-offs in which any trait which benefits young organisms tends to be favored above those which benefit old organisms. In particular, a trait which favors young organisms *even at the expense* of the survival of the same organisms when old, will still tend to be favored by natural selection.

An important instance of a trade-off which benefits the young organism but harms the old organism is found in the optimization of investments in long-term maintenance and repair of bodily cells and tissues. Maintenance and repair, through processes like DNA repair or cell renewal, require some expenditure of metabolic energy. This expenditure is necessarily at the expense of resources that otherwise could be used for activities which directly enhance Darwinian fitness like growth and reproduction. Therefore, the organism should not invest greater effort in maintenance and repair than is necessary to remain in reasonably sound condition through its natural expectation of life in the wild environment. This concept provides the basis of the *disposable soma* theory of aging.

DISPOSABLE SOMA THEORY

The disposable soma theory predicts that aging results from the gradual accumulation of unrepaired faults in the cells and tissues of the body, through the evolved limitation in the investments in maintenance and repair. Different species evolve to have different life expectancies because they are subject to different levels of environmental hazard. For example, a species subject to high levels of environmental hazard, such as a mouse, should not invest much effort in maintenance and repair but should instead invest in a high rate of reproduction. A species subject to low levels of environmental hazard, like an elephant, should do the reverse. Humans have evolved a series of adaptations, driven primarily through evolution of a large brain, that have dramatically reduced the level of environmental hazard. These presumably explain why humans have the longest life span of all mammals. Nevertheless, the rapid reductions in mortality that have occurred within just a few human generations have outstripped any immediate possibility of evolutionary adjustment to the determinants of our longevity. The maintenance and repair mechanisms in our bodies evolved at a time when life expectancy at birth was forty years or less.

Current research on the genetics of longevity and on the mechanisms of aging and age-related diseases is focused on understanding the kinds of damage that affect the cells and tissues of the body and on the cell maintenance and stress response systems that protect us. One of the clear messages from the disposable soma theory is that there is no *single* mechanism of aging. A large number of maintenance and repair systems collectively provide the network of cellular defense mechanisms (Kirkwood & Franceschi 1992) and it is predicted that all of these participate to some degree in longevity assurance. The evolved limitation in the investments in maintenance and repair, as predicted by the disposable soma theory, means these mechanisms will be insufficient to prevent the gradual accumulation of damage. Thus it is to be expected that damage will accumulate in various forms, e.g., somatic mutations in DNA, accumulation of altered or aberrant proteins, oxidative damage to membranes, etc. There is experimental evidence for accumulation of each of these types of damage in a wide range of organisms as they age. Furthermore, comparative studies have shown that long-lived animals accumulate damage at slower rates than short-lived animals and have higher tolerance for intrinsic and extrinsic stressors, such as free radicals. Certain animal models, such as fruit flies and nematodes, are readily amenable to study of mutants with altered life spans, to transgenic manipulations affecting normal aging, and to artificial selection for increased longevity; these studies have consistently shown that increased cell maintenance and stress resistance is associated with increased life expectancy, and vice versa.

INTERACTIVE BIOLOGY OF AGING

An important area of research concerns the integrative biology of aging, namely, the interactions between different mechanisms of aging as they affect individual cells, and how age changes in cells affect the functional integrity of tissues and organs. Multiple biochemical mechanisms contribute to cell aging, and cellular age changes have complex effects on the functions of tissues and organs. An example is the effect of aging on stem cells of the intestinal wall. The wall of the intestine is one of the most highly proliferative tissues of the body, which is continually renewed from the division of the tissue stem cells. Stem cells occupy a special position at the base of small pocket-like structures, called crypts, embedded in the wall of the intestine and they divide periodically to give rise to specialized daughter cells which undergo rapid cell divisions as they migrate outwards along narrow finger-like projections, called *villi*, from which they are eventually lost into the gut. The maintenance of the integrity of the relatively small number of stem cells is essential for the long-term functional integrity of the entire gut wall. Because of this critical requirement, stem cells are unusually sensitive to damage. When exposed to stressors, such as low-dose irradiation, damaged stem cells undergo programmed cell death, or suicide (also known as apoptosis), and their place is normally taken by the daughter cells formed from the division of an undamaged stem cell. This programmed cell death is thought to benefit the organism as a form of stringent quality assurance to protect against the accumulation of damaged stem cells, which may become malignant. However, during aging it may happen that many of the stem cells accumulate low levels of intrinsic damage, and this could explain some of the deleterious changes seen in aged tissue. Recent evidence shows that stem cells in old animals are more likely to undergo programmed cell death following low-dose irradiation than stem cells in young and middle-aged animals, and that stem cells in old animals are also less able to regenerate tissue after damage (Martin et al. 1998; and unpublished data). It is an interesting question whether the function of the intestinal wall in old age might be improved by altering a stem cell's propensity for programmed cell death (and in which direction?), or by identifying and intervening in the underlying mechanisms responsible for the accumulation of intrinsic damage. This example illustrates the need for fundamental studies of aging mechanisms before we can be confident of attempting interventions in humans.

HEREDITY AND AGING

Another important area of advance is in the study of genetics of human longevity. It is clear from twin studies that there is a significant heritable component to human longevity (McGue et al. 1993). Some early progress is now being made in identifying some of the genetic factors which may be involved, and as the human genome project develops, this work is likely to accelerate (Kirkwood 1997). Whether such findings will ever result in targeted genetic modifications aimed at increasing life expectancy is far from clear. It is more likely, at least in the near future, that genome research will lead to the further identification of risk alleles for late life diseases, such as has already been done for Alzheimer's disease and certain forms of cancer. If such discoveries are coupled with the development of effective drug therapies or prophylaxis, or with lifestyle interventions, such as nutrition and exercise, they will be of considerable benefit.

CONCLUSION

The prospects for a significant increase in our knowledge of the biological determinants of longevity have never been brighter. Nevertheless, it would be unrealistic to expect a process as complex as aging to yield quick and easy solutions. There is an urgency in pursuing basic research on the science of human aging if a timely scientific basis for extending quality of life in old age is to be developed. (Holliday 1996).

REFERENCES

1. Holliday, R. 1996. The urgency of research on aging. Bio Essays 18, 89-90.

2. Kirkwood, T.B.L.1997. Genetics and the future of human longevity. Journal of the Royal College of Physicians of London 31, 669-673.

3. Kirkwood, T.B.L. & Franceschi, C. 1992. Is aging as complex as it would appear? Annals of the New York Academy of Sciences 663, 412-417.

4. Kirkwood, T.B.L. & Rose, M.R. 1991. Evolution of senescence: late survival sacrificed for reproduction. Philosophical Transactions of the Royal Society, London B332, 15-24.

5. McGue, M., Vaupel, J.W., Holm, N. & Harvald, B. 1993. Longevity is moderately heritable in a sample of Danish twins born 1870-1880. Journal of Gerontology 48, B237-244.

6. Martin, K., Kirkwood, T.B.L. & Potten, C.S. 1998. Age changes in stem cells of murine small intestinal crypts. Experimental Cell Research, 240, in press.

SPECIFIC SOCIAL RESPONSES TO AGE-RELATED DYSFUNCTION

Alvar Svanborg

University of Illinois at Chicago
Department of Medicine
8400 South Wood Street
Chicago, IL 60612-7323
USA

INTRODUCTION

This presentation will focus on the possibilities of distinguishing the manifestations of aging from symptoms of disease prevalent in older persons, when disability and morbidity become common.

VARIABILITY IN LONGEVITY

Are differences in longevity between nations, and between segments within a given population, caused not only by differences in incidence/prevalence and treatment of disease, but also by differences in the rate and functional consequences of aging itself? The goal of avoiding stagnation accompanying increasing longevity should then imply both attempts to improve health and postpone aging-related functional decline.

Not only genetic but also external factors influence some functional consequences of aging. Consequently, along with extension of longevity, individual and societal contributions and interventions would stimulate progress and minimize stagnation due to aging per se. Measures to prolong life with productive vitality might, however, be different in different populations, environments and social settings.

It should also be realized that some aging-related changes in functions seem to be mainly genetically determined, and might not, according to current knowledge, be influenced by external factors. When "trainability" of certain organ functions exists for some functions but not for others, risks of side effects by activation of the whole individual have to be considered. This illustrates the need for teachers, coaches, psychologists and medical personnel (e.g., physical therapists)

Longevity and Quality of Life, Edited by Butler and Jasmin
Kluwer Academic / Plenum Publishers, New York, 2000.

with special training in gerontology/geriatrics. In general, more knowledge about aging has to be introduced to young and old in our societies.

For several decades, pediatricians have taken responsibility not only for treating and preventing sickness in children, but also for understanding and promoting their growth and maturation. Adult medicine has, on the other hand, only recently focused on trying to understand the aging process per se, and on discovering ways to postpone decline in function and vitality. One obvious reason was the problem of differentiating between normal aging and the diseases of longevity. For example, the older we become the higher the risks of becoming tired and short of breath. Certain changes in laboratory findings might be caused by aging, however, almost identical changes might reflect disease. Such similarities between normal aging and symptoms of disease have constituted–and continue to constitute–risks for both under- and over-diagnosis as well as under- and over-treatment, especially in frail older persons.

Except for observations in animals with shorter life spans than humans, crucial scientific information about aging in general, and specifically about aging in humans, has been obtained from longitudinal retrospective and prospective studies in population samples, allowing generalizations to whole populations, or at least to definable parts of populations. It is impossible to obtain a scientifically acceptable evaluation of an old person's health through a single detailed examination. Longitudinal follow-ups are needed. Toward this end, non-invasive examination techniques developed in recent decades have been a great asset, allowing detailed morphological, physiological and sometimes organ specific biochemical analyses without risks or inconvenience to the participant.

Such longitudinal studies have markedly improved diagnostic criteria in old people, and have also enabled the identification of manifestations of aging per se in frail older persons. The information constitutes the basis necessary for the identification of factors influencing normal aging as well as available means to postpone aging-related functional decline. In other words, science has reached a level of understanding that aging is influenced by exogenous factors, some identified as good, others as bad. And some of those factors that externally influence aging could be applied in intervention programs.

Many of the improvements that have been made in diagnostic and treatment criteria for diseases common in old people could be applied to most populations in the world. But we lack population studies that would allow us to make global generalizations about how people grow old. One of the many prerequisites for studies of population aging is a reliable census register in every country, obviously containing the birth dates of its citizens. Many countries need more information about their older peoples' health status and level of vitality. Furthermore, to expand the knowledge about exogenous influences on aging and vitality, a comparison of "contrast groups," that is, of populations with major differences in living circumstances, would be important not only for developing populations, but also for the industrialized world. International support for longitudinal population studies in developing country and studies allowing comparisons with industrialized countries where data already exist, would be essential in order to widen our understanding of aging per se, and also to plan interventions aimed at minimizing the risk of decline with the increase in longevity in many developing countries.

A worldwide perspective reveals that the greatest variations in longevity (both increases and decreases) exist between the developing nations and the industrialized nations of the world. However, it should be noted that variances in longevity also exists between nations with very similar socioeconomic standards, living conditions and availability of social support and medical services, for example, the populations in the Nordic countries. It is well known that the diagnosis on a death certificate is often inaccurate. In many countries, that is less true for malignancy than other diseases. Official causes of death in Denmark, Finland, Norway and Sweden, show a similar pattern. They are countries with relatively comparable medical organizations and rules governing documentation of cause of death. This indicates that factors other than morbidity and the availability of medical service and care, account for the differences in longevity in these populations. A tentative but probable conclusion would be that people grow old at different rates in these countries.

Even more marked differences in longevity exist between other European countries, e.g., Hungary and Sweden. Hungary represents one of the European countries in which longevity for males has been reported to show a declining trend for many years, widening the gap in longevity in comparison with many other countries. The reasons for such ongoing negative versus positive trends in change of life expectancy in so many other countries is not clear, and is probably multi-factorial. Would anyone insist that such a decline in longevity implies social progress? Research is needed to explore the cause of differences not only between developing and developed countries, but also within populations.

WOMEN AND LONGEVITY

In general, nowadays females live longer than males. A phenomenon observed in the Swedish population since the early eighteenth century is that they already had a greater life expectancy after age 65 and lived longer than men when they became old. Even before puerperal fever, which influenced humankind's total life span, was virtually eradicated, it was noted that women exhibited a biological superiority over males with regard to further life expectancy over 65.

Increase in longevity accelerates faster in females in many populations. If this trend in sex difference persists the "risk" that females will contribute to a feared, but not proven longevity-induced social stagnation would, thus, theoretically be higher.

FACTORS INFLUENCING LONGEVITY

Interventions to Extend and Enhance Longevity

Both more detailed research of specific organ functions, and epidemiological research of human population samples have shown that our lifestyle, the environment, and the availability and usage of medical care have a profound impact on the rate and functional consequences of at least certain manifestations of aging. Unfortunately, such studies have hitherto been unable to demonstrate that such exogenous factors would be beneficial to aging in general. For certain crucial manifestations of aging, genetic factors seem to dominate. We cannot do much to postpone aging-related decline in functional performance for which those factors play an important role. The goal of this presentation will be limited to an attempt to exemplify and illustrate some controllable and some non-controllable functions. The aim is to elucidate how further extension of longevity, and especially an extension of further life expectancy when we already are old, would encourage or impede societal progress or stagnation

Several studies directly or indirectly illustrate how our lifestyle might influence muscle strength and quality of muscle function in our body. Most studies have focused on striated muscles; fewer have studied the heart muscle and fewer still smooth muscles. Systematic physical training has been reported to increase striated muscle strength in persons above the age of 80.

These observations are one reason for an attitude change in geriatric medicine toward a much more optimistic view of possible outcomes of reactivation programs than those which previously dominated treatment programs for old and frail people. There is also evidence that the strength of the heart muscle can be improved by careful systematic training. Training can increase the amount of blood expelled per heart beat, i.e., the stroke volume. Such studies also indicate that in younger years a low level of physical activity implies a more marked physical response to the declines caused by aging. Even blood pressure level, commonly increasing during life and up to age 70-75, is influenced by several exogenous factors such as body weight, and physical activity habits. Reports have shown that aging-related change in blood pressure can vary between generations born with rather short time intervals. Blood pressure increase is known as a risk factor for many cardiovascular and cerebral disasters often seen in older persons. Certain changes in lifestyle can thus be expected to contribute to a delay in aging-related cardiovascular function disturbances and disasters in the old, and minimize the risk of social stagnation caused by a future predicted increase in longevity.

Numerous studies have also shown that exogenous factors influence not only the building up of the amount and stability of bones, but also the rate of decline of skeletal mass. Exogenous factors also partly influence the development of the clinical syndrome of osteoporosis which commonly causes fractures, especially in women after the menopause. Some of these fractures are life-threatening in old persons. Several exogenous factors influence the development of osteoporosis, such as nutritional components like vitamin D, calcium, certain foods and components similar to sex hormones, etc. During recent years the importance of the quantity and quality of physical "weight bearing," for example walking, or more specifically designed forms of physical activity, are other examples of lifestyle factors which can postpone the rate of decline of skeletal stability during aging. Both tobacco smoking and alcohol abuse accelerates the development of osteoporosis.

Falls and resultant fractures are dependent on muscle strength and the ability of the extremity muscles to react rapidly. Balance is dependent also on many other organ functions, such as the central and peripheral nervous system, visual acuity, hearing, general stiffness of the body, the existence of joint disorders, sudden fall in blood pressure when standing up, etc. Although balance defects can be due to non-reparable dysfunction, in general clinical experience shows that much more can be done to help old people with balance problems, for example, after periods of inactivity, than usually considered.

These examples of how exogenous factors might postpone aging-related functional decline indicate that certain reserves are available and can sometimes still be mobilized at advanced age. The probability that these reserves are shrinking with advancing years is obvious, even though it has been demonstrated that striated muscles can still be trained in persons aged 85-90 years.

Many older people don't receive enough help in regaining functional performance after events like an acute disease, loss of a spouse, periods of unwanted social isolation and inactivity, etc. In medicine it is common that a disease is treated, but the patient is sent home before she has reached a state where her own reserves allow her to recover strength and function. The author calls this a lack of "reactivation," instead of the generally used term "rehabilitation" because of the special geriatrically-dominated needs these old frail people have. Much more can be done to help through reactivation of old frail people than is currently provided. And in a cost-benefit analysis, doing more to help an individual experiencing a life-threatening event might result in savings for society.

Functional Declines

There are also aging-related functional declines in which exogenous postponing influences have not been documented. With increasing age many organs and tissues become increasingly stiff. Such a decrease in compliance can be caused by changes in the cells central to the organ function. A common cause is a stiffening of the connective tissue which is important for the structure and function of many organs, such as muscles, heart and vessel walls, skin, tendon and joint capsules.

Although some reports have demonstrated a higher elasticity in connective tissue components in well-trained older persons, and although physical activity is proven to stimulate metabolism in components like collagen in connective tissue in certain organs even in an old individual, the possible existence of "trainability" that really would imply a significant postponement of tissue stiffness through intervention in already old humans has yet to be proven.

Another example of functional declines that seem to be mainly influenced by genetic factors, and presumably not influenced by factors such as change in lifestyle and living conditions, concern the neuroconductive system of the heart. The number of pacemaker cells in the "initiator" of cardiac rhythm, the sinus node in the right atrium of the heart, is reported to decline during aging. The number of neurofibers transporting impulses from a similar node at the border between the atria and the chambers, the atrioventricular node out to the chamber walls are also diminishing in numbers. Such changes in the neuroconductive system in the heart will increase the probability during aging of rhythm disturbances. And they have to be considered in training and reactivation of older persons.

CONCLUSION

Thus, more possibilities exist to delay aging-related functional decline than previously generally considered. That is the good news. But there are also crucial organ and organ system functions for which we lack proof that meaningful interventions aimed at a postponement exist. This makes it difficult to predict to what extent further increases in longevity in the future will result in societal progress through the existence of more productive and experienced people, or the opposite, causing stagnation or even a decline in the average living standard. One can always hope that more old, experienced and still productive members of our societies will foster social progress, even though aging-related "dysfunctions" might cause some burdens and require changes in the societal structure and function.

The present paper has focused on aging-related dysfunction, not on changes in morbidity patterns in graying nations, and on the present era of medico-technical rapid advancements influencing health and vitality, even in the very old.

IN YEAR 2000: AGING, THE FORGOTTEN REVOLUTION WILL DHEA HELP US?

Etienne-Emile Baulieu

INSERM
80 Rue de General Leclerc
Le Kremlin Bicetre - Cedex 94276
France

INTRODUCTION

The year 2000 holds no special meaning for me, but it does serve to remind us, with the help of a few festive bursts, of the uniqueness of *Homo sapiens*, which, although not particularly virtuous, is a really intelligent species!

A unique and fragile destiny in the midst of a universe of overwhelming forces: a mere flick of a celestial finger could wipe mankind and all living forms off the face of our planet Earth. In the meantime, life plods on, strangely impervious to the wind, water and vibrations that erode mineral constructions. Life both implies and resides in the cycle of reproduction and death and, by constantly adapting. Its continuity is assured thanks to the Darwinian process of selection, which singles out those species capable of advancing further along the evolutionary highway. Is man the most complete of living creatures? Who knows?

Two revolutions are currently taking place with no heads rolling or rivers turned to red. Each and every one of us is affected and it's happening very fast.

The first revolution has already been taken for granted. In the wake of electronic and information technology (IT) leaps and bounds, the communications revolution is changing and will continue to change the way we work, our lifestyles and our relationships with others. Not nearly enough interest is being paid to this revolution, despite the numerous benchmark studies and research that are available.

The other revolution is virtually unknown, or at least ignored. We are talking about the longevity revolution, which is striding ahead, albeit discreetly, not making the sort of fuss that such a radical change to human society, which challenges our current lifestyles and the socioeconomic issues of all societies, deserves to make. With the dawning of the 21st century, it will not be long before

Longevity and Quality of Life, Edited by Butler and Jasmin
Kluwer Academic / Plenum Publishers, New York, 2000.

the youngest of the baby boomers turn into senior citizens, around 2030, and trigger a demographic explosion of aging populations.

In terms of longevity, humankind has already achieved a great deal. Unlike most species, survival continues for a long time after the reproductive period (limited to the age of 50 in women) has come to an end. Over the last century, the average life expectancy of men and women in industrialized countries has been extended by 25 years. This is none-too-surprising given that women in childbirth, premature babies and old people with pneumonia no longer die in large numbers, and many other illnesses have either been eradicated or are curable. Progress is spectacular and will be even more so in developing countries, where hygiene and standards of living are still rudimentary. Even more surprising, life expectancy continues to rise in our "advanced" nations, even though no remarkable changes, either in medicine or in the distribution of wealth between the poor and the affluent, have occurred.

For example, since 1970, life expectancy in France has been increasing by almost one year every four years, and there is no reason to think that it will not continue, both here and in the developing countries (about which it will be impossible to make any definitive judgments until underdevelopment has been rectified).

This increase in longevity has a direct impact on population size: demographic increases are the result not only of higher birth rates, but also of deaths which have been forstalled. Birth control is becoming increasingly indispensable, since it is one of the only means of ensuring that population growth remains compatible with the wealth of a community, and that lifestyles can be maintained over future generations. Even when the scientific academies of a number of countries gathered at an international conference to discuss a document entitled, "For sustainable consumption," they did not broach the subject of increasing life expectancy. What are all the old people going to do? How can they be useful to society? How is the financial and economic equilibrium of this new age pyramid to be ensured?

Let's go back to the basic issue. It is generally admitted that each species has a fairly well-defined life expectancy, as if this were a genetically encoded and irrevocable fact. This concept is borne out both by genetic experiments, in particular on very simple life forms such as the thread worm *Caenorhabditis elegans*, and family trees like that of Jeanne Calmant's. Yet, studies conducted on fruit flies and rodents show that cutting down on their calorie intake increases their life expectancy; and other research has proved that the life span of fish can change as a function of water temperature. Nothing so far proves that a central biological clock exists in man to regulate with precision the span of a human life. Depending on the individual case, organs, regulatory systems or metabolism are modified over time, eventually producing an impact that is critical to the whole organism. The individual's environment in the broad sense (habitat, nutrition, physical and mental activities) would appear to play a decisive role in speeding up or slowing down the "wear and tear" of cell mechanisms, as well as in preventing cells from being renewed and thereby providing vital elements to the survival of the body as a whole.

In other words, we can expect to live a long and healthy life with cerebral functioning intact provided that we are aware of the products and conditions necessary to maintain our basic biological mechanisms, both general and specific, for each organ or metabolic function. However, very little biological and medical research is being done to make those extra 10, 20 or 30 years that await us in the 21st century healthy ones.

But . . . are individuals really ready to live longer? Is society prepared to make this extended existence a profitable one, both individually and collectively? A great many changes must occur in response to the simple fact that longevity is increasing. They will probably be combined with the changes brought about by the information technology revolution underway, which is completely transforming working conditions and, consequently, each individual's entire life. Life's successive stages are:

1) education, 2) working life, which is more or less suffered, since it is considered a relatively "short-term" necessity, 3) retirement, and exclusion from society. These stages will be modified and this apparently ingrained pattern will be turned on its head sooner than we imagine. These extra years can only be happy and beneficial for everyone if two sets of problems have first been resolved:

1) First and foremost, individuals must maintain good physical health with a minimum of handicaps so as to be able to enjoy life and participate in family and community activities. Particular attention must be paid to the central nervous system, since neurologic abnormalities can be disastrous for the individual and the immediate community. Functions must not be allowed to regress, so that everyone can pass on the souvenirs of a lifetime of experience, thereby benefitting the whole of society. Traditional "African wisdom" places greater value on the experience of elders than do our societies.
2) New ways of living have to be created so that elderly persons can remain integrated in society. Certain useful activities require less physical strength, but more experience. The barrier represented by retirement and the exclusion of people by dint of their date of birth, is belittling and incapacitating to the point of transforming personality and self-perception.

Making this new stage of life possible will have serious consequences, and will call into question our entire notion of social functioning: at work, will it still be possible to promote people based on their length of service? How can responsibilities be shared between young and old? How can jobs be adapted to suit individual capabilities, skills and changing ambitions? In what ways can reintegrating older workers who so wish be beneficial to society? Such reintegration will only be sought by a minority, since the idea that one deserves to rest after a hard-working life is deeply ingrained in society, and since efforts to improve the health of aging persons have been so inadequate to date. Still, hope should not be abandoned. The homogenization of society, in which most senior years become a part of active life, will perhaps be representative of the coming millennium.

ABOUT DHEA

Aging is accompanied by a progressive and nearly universal decline of the physiological functions involved in maintaining the body's composition and homeostasis during action, emotion and, eventually, response to aggression. Is the aging-related decrease of DHEA (dehydroepiandrosterone prasterone) [1], the steroid at the highest concentration in the blood of men and women, an important component or even a determinant with respect to the difficulties and diseases so frequent in aging? Is compensating for its decrease not only harmless, but even helpful after a certain age? This is the challenge confronting us, and it is indeed a difficult one to resolve.

BIOCHEMICAL ENDOCRINOLOGY

DHEA is a steroid, retaining cholesterol's Δ5-6 double bond and its esterifiable 3b-hydroxyl group. The side-chain cleavage transforming the C27 cholesterol to the C19 DHEA involves, in all probability, several cytochromes P450 with specific hydroxylase (oxygenase)-desmolase activities and mostly expressed in steroidogenic cells of the adrenals and gonads. (Some controversy remains, evoked by S. Lieberman et al. (1). The steroid was first described by A. Butenandt in 1934 after acid hydrolysis of urine, which splits off its conjugated sulfate ester. Sulfoconjugation, one of the most common modes of inactivation and elimination of hydroxylated substances, including hormonal metabolites, is mainly performed in the liver. However, at the end of the 1950s, working with a large adrenal tumor from a patient with an extremely high level of DHEA circulating in the blood as a sulfate ester, I was unable to detect the unconjugated steroid in the tumor (2). In the framework of studies carried out in M.F. Jayle's laboratory, I had developed extraction and chromatographic procedures permitting the identification of "conjugated" steroids. Thus, I could characterize DHEA sulfate (DHEAS) in the tumoral tissue. Subsequently, the synthesis of DHEAS and its physiological secretion by the adrenal cortex in healthy men and women was demonstrated. S. Lieberman , R. Vande Wiele, and S.&J. Tait, and their colleagues incorporated this result into their studies on the production and interconversion of androgens, and sulfoconjugation of steroids were no longer considered to be a simple hepatic detoxification mechanism [2].

Free DHEA and DHEAS are metabolically inter-convertible (by phospho-adenosine-phosphosulfate [PAPS] -dependent sulfotransferase for conjugation and sulfatase for hydrolysis, both found in many tissues). Using mostly radio-isotopic tracer methods and calculations with a minimum of arbitrary assumptions, the half-life of DHEA is estimated (similarly to that of most steroid hormones) to be 15-30 min., with a Metabolic Clearance Rate (MCR) of ~2,000 l/day, while the half-life of DHEAS is much longer, 7-10 hours, and the MCR is low, 5-20 l/day. The differences in plasma concentrations logically follow: a few ng/ml (2 to 4) for DHEA (~ 1.10^{-8} moles/l) and several μg/ml (2 to 6) for DHEAS (~ 1.10^{-5} moles/l) in young adults (20-35 years), ~ 10-20 percent more in men than in women. In healthy people, 60-90 years old, the concentration is 0.3-2μg/ml within the 80-85 percent confidence limits. The DHEA/DHEAS ratio in the blood is reported to be higher in women [3]. The order of magnitude of adrenal secretory rates is, in young adults, ~ 4 mg/day for free DHEA and 25 mg/day for DHEAS.

Physiologically, the concentration in the blood of DHEA oscillates, coincident with cortisol and consistently with the response of adrenal DHEA secretion to ACTH, but there is no feedback control of DHEA secretion at the hypothalamo-pituitary. ACTH also stimulates adrenal DHEAS synthesis, but due to its low MCR, the concentration of the steroid sulfate remains at approximately the same level 24 hours a day under normal circumstances. In the blood, neither DHEA nor DHEAS bind significantly to Sex Steroid Binding Plasma Protein (SBP) or Sex Hormone Binding Globulin (SHBG), and the binding of DHEAS to albumin seems nonspecific. Very little of plasma DHEA(S) appears to be of testicular or ovarian origin.

In monkeys (great apes, baboons, macaques), there is some DHEA in the blood, but much less DHEAS than in humans. In fact, in most if not all laboratory animals (particularly rodents), the DHEA(S) concentrations are so low that it is impossible to detect a significant decrease with age as that found in humans. It is important to take the human species specificity of DHEA secretion and concentration in the blood into consideration when planning studies on DHEA(S), particularly with respect to aging.

Secreted DHEA(S) are metabolized by a number of enzymes very widely distributed in the body. Besides the DHEA(S) interconversion, there is formation of androgens (thus the inappropriate nickname, "weak androgen", attributed to DHEA(S), which is not in itself androgenic and does not bind to androgen receptors), such as testosterone (T) and 5α-dihydrotestosterone (DHT), while Δ4-androstenedione (Δ4-A), abundantly formed, is also inactive per se but transformable into T and DHT. These active steroids are further metabolized, mostly to conjugated, reduced steroids which are inactive and for excretion. The complexity is increased because T and Δ4-A can themselves be "aromatized" to give estrogens. In addition, DHEA may be metabolically transformed to the weakly estrogenic Δ5-androstene-3 β,17βdiol and other products such as 16 α-, and 7 α ~ β -oxygenated DHEA derivatives of yet poorly defined function.

When DHEA is administered, orally or parenterally (including transdermal route), metabolites are formed in different proportions according to the administration route. After oral administration, DHEA is largely absorbed and converted to DHEAS in the hepato-splanchnic system and the blood concentration of the latter increases rapidly (3). There is no rationale for oral administration of DHEAS, as it could be largely hydrolyzed in the acidic medium of the stomach.

The metabolism of DHEA into potentially active sex steroids may occur in many cells containing androgen or estrogen receptors; i.e., adipose tissue, bone, muscle, breast, prostate, skin, brain etc., and particularly the liver, where it is quantitatively important, and from which part of these active DHEA metabolites may be released into the circulation and thus reach target tissues.

DHEA(S) DURING DEVELOPMENT AND AGING

The ontogenic pattern of DHEA(S) secretion in humans is unique. Synthesized in large amounts by the fetal adrenals, it is the source of estrogens made in the placenta directly or after partial subsequent 16 α -hydroxylation in the fetal liver. At birth, the "fetal zone" of the adrenals regresses, and the reticularis layer only starts to synthesize DHEA(S) again at 5-7 years of age in both sexes, defining "adrenarche." Thereafter, there is an accelerated increase through the period of puberty, and the maximum blood concentration is reached during the third decade of life. Then starts an ineluctable decline, cross-sectional studies indicating a decay of ~ 2 percent /y in the blood level and thus leaving a residual value of ~ 10-20 percent during the 8th-9th decades of life (4, 5).

This decrease is not due to a change in the metabolism of DHEA(S), but to a diminished adrenal secretory rate. There is no demonstration of an adrenal androgen stimulating factor which could be involved, and no decrease of ACTH may explain it. Growth Hormone (GH) and Insulin-like Growth Factor-1 (IGF-1), which decline with age, do no affect DHEA(S) levels and the possibility of a physiological interaction between high insulinemia (which tends to increase with age) and decrease of blood DHEAS is still uncertain, as indicated by J. Nestler at recent meetings [4]. Challenge tests with ACTH indicate that in elderly subjects the DHEA(S) response is decreased, unlike that of cortisol secretion which is maintained. There is apparently a specific, but unexplained, defect in the desmolase activity of the cytochrome P450c17 in the adrenal reticularis zona. The increase of the cortisol/DHEA(S) ratio in the blood is possibly partly responsible for the catabolic state which develops during aging (just the opposite of that which occurs in fetal life and adolescence, when DHEA(S) is very high).

Interestingly, among human beings of the same age and apparently in good health, there is a high inter-individual variability in the concentration of DHEAS (up to threefold or more), and therefore the practical interest of DHEAS determinations for clinical use has been questioned, contrary to those of cortisol, which is routinely assayed in internal medicine to assess adrenal function. However, when in longitudinal studies DHEAS and cortisol values are compared (in samples carefully obtained between 7.00 and 8.00 a.m. to minimize circadian cortisol variations), the stability over time of the ranking of normal subjects with respect to blood steroid concentrations is higher for DHEAS than for cortisol, whether measurements are made at six month intervals in old women or every 2-3 days in young men (9). In our longitudinal studies and those of N. Orentreich (5), intra-subject variations of DHEAS are greater than those of cortisol in absolute values, but are stable relative to the corresponding inter-subject variability, and this is in agreement with a genetic component having a significant influence, as has previously been suggested by L.N. Parker et al. (10). In other words, DHEA(S) concentration is a highly specific individual marker.

However, it remains that a basic explanation for both the adrenarche increase and the age-related decrease of DHEA(S) secretion in human beings is lacking, although it is one of, if not the, best measurable indices of chronological aging. We have observed, in a group of elderly women (≥70 years), a relative longitudinal decline of DHEAS throughout the study that is greater than when cross-sectional analysis is carried out (9). This result may lead to the hypothesis that higher DHEAS levels are associated with longer survival in this group of women, so that there is a continuous selection of individuals with higher levels. However, even if there is correlation between plasma albumin and DHEAS concentrations in aging (A. Vermeulen), and even if experiments suggest antioxidant properties of DHEAS at physiological concentrations, no clear data concerning the predictive value of DHEAS levels for length of life has yet appeared in epidemiological studies. At best, DHEA can only be but one constituent among others influencing longevity, and there is no evidence for a vital function depending exclusively on DHEA(S). (Patients without functioning adrenals survive with the help of corticosteroids without adding DHEA.)

Effects of DHEA

Until recently, no specific molecular or cellular target of DHEA(S) had been described, and no experiment has clearly demonstrated specific binding of DHEA(S) by a protein/receptor nor any measurable modulatory effects on a given tissue or cellular system, whether tested *in vivo* or *in vitro*. However, in the last few years, striking effects have been obtained, mostly in animals, with high doses of DHEA which would generate excessive quantities of steroidal sex hormones if correspondingly applied to people. We indicate below the few data available for humans as they are related to experimental results (6-8).

Very large doses of DHEA have demonstrated anti-tumoral effects (A. Schwartz et al.), provoking prevention and/or regression in spontaneous or chemically-induced skin and colon cancers, in specific strains of rodents. Independently, a few reports have suggested low production of DHEA(S) in women at risk of, or having breast cancer.

The preventive and/or therapeutic activity of DHEA has been reported in animals with diabetes of genetic origin or induced by streptozotocin, and a reduction toward "younger" levels of the elevated blood insulin has been obtained in old rats. There are also experiments assessing the anti-obese effect of DHEA in genetically obese rats, suggesting the induction of caloric inefficiency, sometimes associated with extension of the length of life. In men, low DHEAS levels have been observed in obese patients, and oral administration of 1,600 mg/day of DHEA for a few weeks in normal individuals led to decreases of cholesterol, LDL and body fat, a result not confirmed by studies in obese young men and postmenopausal women.

DHEA dramatically decreases the development of arterial lesions in rabbits on an atherogenic diet. In humans, several reports indicate low DHEA in hypertensive patients, and cases of cardiac allograft vasculopathy. However, in spite of frequent association of risk factors with low levels of DHEA, correlations with extension of atherosclerosis, occurrence of fatal coronary disease and death related to cardiovascular deficiency remain uncertain.

There is also an effect of DHEA in protection against virus development in animal experiments where the compound (and the more active 16 α-bromo-epiandrosterone) inhibited Epstein-Barr virus-induced transformation of lymphocytes, and was protective in two lethal viral infection models (coxsackievirus B4 and herpes simplex type 2 viruses, R. Loria) and bacterial infection (streptococcus faecalis). *In vitro* effects on HIV-I replication in macrophages and lymphocytes have been reported. These effects may involve an immunological mechanism. A protection effect against glucocorticosteroid-induced involution of the thymus also has been observed, as well as attenuation of age-related decrease of immune response to vaccines. Immunosenescence includes the lowering of the immunological defenses with changes in the expression of T and B cells. It has been reported that, with aging, there is an increased production and release of interferon γ (IFN γ), and interleukins IL4, IL6, IL10, and a decrease of IL2, IL3 and GMCSF (R Daynes). Experiments in animals, and some with human

In particular there may be a stimulatory effect on IL2 secretion, prevention of the IL6 increase, and modulation of natural killer (NK) cell differentiation.

The effects of DHEA(S) in the central nervous system are of particular interest. DHEA(S) is a neuroactive neurosteroid in rat brain (11, 12), being an inhibitor of the $GABA_A$ receptor, and a potentiator of NDMA and sigma receptor function. Administration of DHEA decreases the aggressivity of castrated male mice toward females; it also increases retention of learning in rats, and a neurotrophic effect on neurons and glial cells has been described. Recent work has shown that pregnenolone sulfate, considered a physiological precursor of DHEA(S), is selectively decreased in the hippocampus of aged (24 months) rats with deficient cognitive performance and that *in situ* administration of nanogram amounts of pregnenolone sulfate repair (temporarily) the memory deficit (13). Although far from being transferable to the understanding and treatment of memory decrease in aging human beings, these results are still quite encouraging. In humans, very little is known: the levels of DHEA(S)–not necessarily related to brain DHEA(S)–once adjusted for age, are not abnormal in the blood of patients with neurodegenerative diseases (and preliminary treatments with DHEA of Alzheimer's disease patients have not given appreciable results). There are low levels of DHEAS in nursing home patients related to their degree of dependence in activities of daily living (14). A remarkable correlation between blood DHEAS and the expressed feeling of well-being has been observed in a large scale epidemiological study on ≥65 year old people in the South-West of France (Paquid) (15). We do not know the mechanism of such an effect, nor can we demonstrate a cause-effect relationship, but it is certainly consistent with the registered change observed after a few weeks oral administration of 50-100 mg per day of DHEA to aging subjects, where the increase of well-being, both in men and women, was particularly impressive (16). It is possible that the activity of DHEA(S) involves the easy penetration of DHEA through the blood-brain barrier, the interconvertibility of DHEA(S), and possibly action on the brain of steroid metabolites or IGF-1 induced by DHEA.

In addition, low DHEAS levels have been recorded in a number of circumstances, when people are in relatively poor health due to stress (accident, surgery), to immunological disregulation (systemic lupus erythematosus, active rheumatoid arthritis, AIDS, etc.). This apparent lack of specifity of the effects of DHEA(S) can be understood if it puts the body in better condition to resist an aggression via changes in steroid metabolites, IGF-1 production, modulation of cytokines and immunological parameters, and action in the brain. Globally, the mode of action of DHEAS probably involves two series of mechanisms: transformation of DHEA(S) into active sex steroids which promote overall anabolic effects in several tissues, and effects in the central nervous system through various mechanisms, as described above. Parenthetically, there is a demonstrated inhibition by DHEA (not DHEAS) of the enzymatic activity of glucose-6-phosphate dehydrogenase (G6PDH). This inhibition could decrease availability of NADPH for oxygenase activities which favor free radical reactions, and/or diminish DNA synthesis, two effects it would be potentially useful to obtain in aging and cancer processes, respectively. Nevertheless, it would appear that the Ki for G6PDH inhibition is too high for an actual role for DHEA at a physiological concentration.

A natural product, whose activities are not well enough defined, DHEA(S) is not a "drug" steroid without (up to now) a proper selective receptor, and acting in part through metabolites, DHEA(S) is not a typical hormone. Careful, prudent and relatively long-term studies on DHEA administered to humans may progressively assess its safety, delineate its activity of preventing and/or improving a number of pathological conditions, and allow us to look for a possible slow-down of some consequences of normal aging.

The Use of DHEA in Aging

Cancers and atherosclerosis, decrease of immunological responses and brain functions are frequently observed in aging, along with global changes of metabolism (decrease of lean tissues, increase of fat mass, osteoporosis, etc.). Whether it is relevant to augment the low DHEAS blood level in aging people back to the level of young adults in order to correct some of these age-related phenomena remains a challenge. In the absence of an appropriate animal model, there is practically no alternative to human studies, and aging goes on at the same pace for the observed subject and the observing scientists.

Logic argues in favor of oral administration of DHEA at a dose which provides "young" DHEA(S) levels in the blood, and no T/DHT and E2 concentrations superior to those of normal people of 30-40 years of age. Calculations based on production rates, interconversion between DHEA(S) and metabolic studies suggest that "replacement doses" of 25 to 50 mg once daily can fulfill this double requirement.

The published results of S. Yen and colleagues are noticeable in this respect (16). In a six month randomized, placebo controlled, crossover trial with 50 mg of orally administered DHEA there was a remarkable perceived physical and psychological "well-being" effect in both sexes, 82 percent in women, 67 percent in men after three months treatment (and no change of libido), while the placebo effect remained ~10 percent. Importantly, while GHI levels (lowered in aging people) were unchanged (as were IGFBP-3 levels), there was an increase of IGF-1 and a correlated decrease of IGFBP-1 (the ~ 50 percent increase of the IGF-1/IGFBP-1 ratio suggests better access of IGF-1 to tissues). It is well known that many effects of GH are normally mediated through the GH-induced synthesis of the potent metabolic growth factor IGF-1 mostly in the liver, and that the GH-IGF-1 system is decreased in aging. During this relatively short trial, serum lipoprotein, body fat and insulin sensitivity were not modified. In another study, with 100 mg DHEA orally for six months, lean body mass was increased, as well as knee extension/flexion muscle strength. However, if fat body mass was decreased in men, that was not the case in women. A similar gender difference was also observed for the decrease of DHEA(S) by insulin (J. Nestler et al, 7). These differences between men and women have to be studied in detail, as well as the possibly different effects of DHEA in postmenopausal women receiving different types of replacement hormone therapy. The preliminary observations of changing immunological status by low doses of DHEA are consistent with animal studies (increased T cell expression of IL2-R and increased NK cell number). Similar year-long double blind and placebo controlled studies, in people over 60, are being undertaken in Paris, and will assess tolerance (including surveillance of metabolic parameters and target organs such as prostate and breast) and effects on bone, muscle, skin, cerebral functions and immunological systems. However, analyzing the results will be difficult for a variety of reasons. First of all, the chronological age of a subject does not provide an appropriate baseline since each individual, already unique genetically, differs more and more from others with the passage of time, because of the particular avatars of his/her personal life. We are currently unable to correlate DHEAS blood levels with states of health or prediction of length of life. In order to compensate for the decline of DHEA(S) and to record ensuing changes, we believe that the wisest feasible strategy is to administer a replacement dose over a relatively long period of time, avoiding a possible risk of an excessive increase of active sex steroids. From the results we should obtain enough information to set up more research and answer the public's demand for information on DHEA administration versus aging.

It would also be fascinating to study DHEA administration in younger people, i.e., between ages 40-60 years, particularly if they have low DHEAS levels. This will clearly present another challenge to be met when we know more about the effects of DHEA in older individuals [5]. It would be very important to determine if any preventive effects can be obtained under safe conditions.

In any case, if current trials are successful, that is, if activity and safety are demonstrated, we hope the information will be available to everyone, such that its use may remain a choice.

At present, there is only the intention of relieving, for as long as possible, some of the damage due to aging within the framework of research based on physiological data and under medical surveillance. It will take several years to progress from precise and limited observations to a sure, efficacious product that will be beneficial to many people. We may at the same time learn the physiological effects and therapeutic role of DHEA(S) (17).

Endnotes

1. Even though a potential precursor of some active steroids, DHEA is not the so-called "mother hormone" as some scientists, more tempted by the promotion of a flashy expression than by regular and systematic studies and trials, have imprudently (and wrongly) called it. Seen even under the best light, as an error in good faith, presenting an erroneous concept to a public clearly desirous of better health sooner or later leads to deception and brings discredit on science.

2. Other DHEA conjugates have been described: fatty acid esters of DHEA ("lipoidal ") at a low concentration in the blood [_ 2ng of DHEA equivalent/ml plasma in young adults, incorporated in HDL and LDL, and of unknown mode of action in tissues, and the glucuronoside probably of only excretory fate.

3. In practical terms, I argue against the significance of blood DHEA measurements in most clinical circumstances, since its concentration is more than two orders of magnitude lower than that of DHEAS, and the latter may inadvertently be partially hydrolysed to free DHEA during routine manipulations, resulting in an artefactual increase of the recorded values of endogenously free DHEA.

4. Several authors have important papers which are not individually referred in the very short list of references. The interested readers will find their articles in ref. 6, 7, 8.

5. A double blind, placebo controlled study of one-year daily oral administration of DHEA (50 mg/d) to 280 60-80 years old individuals will be completed in Paris by mid 1999 (conducted with F. Forette, S. Legrain, V. Faucounau and coll., Assistance Publique-Hôpitaux de Paris).

ACKNOWLEDGMENTS

I would like to thank my colleagues for their collaboration, INSERM for long term support, and the Mathers Foundation for its help. Dr Krzysztof Rajkowski, Corinne Legris and Jean-Claude Lambert gave me editorial assistance.

REFERENCES

1. Prasad VVK, Raju Vegesna S, Welch M, Lieberman S. 1994 Precursors of the neurosteroids. Proc Natl Acad Sci USA 91:3220-3223.

2. Baulieu EE, Corpéchot C, Dray F, Emiliozzi R, Lebeau MC, Mauvais-Jarvis P, Robel P. 1965 An adrenal-secreted "androgen": dehydroepiandrosterone sulfate. Its metabolism and a tentative generalization on the metabolism of other steroid conjugates in man. Rec Prog Horm Res 21:411-500.

3. Longcope C. 1995 The metabolism of DHEA. In "Dehydroepiandrosterone (DHEA) and Aging". Ann NY Acad Sci 774:143-148.

4. Migeon CJ, Keller AR, Lawrence B, Shepard TH. 1957 Dehydroepiandrosterone and androsterone levels in human plasma. Effect of age and sex, day-to-day and diurnal variations. J Clin Endocrinol Metab 17: 1051-1062.

5. Orentreich N, Brind JL, Vogelman JH, Andres R, Baldwin H. 1992. Long-term longitudinal measurements of plasma dehydroepiandrosterone sulfate in normal men. J Clin Endocrinol Metab 75:1002-1004.

6. Kalimi M, Regelson W. (Eds). 1990 The biological role of dehydroepiandrosterone (DHEA). New York: Walter de Gruyter.

7. Bellino FL, Daynes RA, Hornsby PJ, Lauvrin DH, Nestler JE (Eds). 1995 Dehydroepiandrosterone (DHEA) and aging. Ann NY Acad Sci 774.

8. Labrie F (Ed). 1995 DHEA transformation into androgens and estrogens in target tissues: intracrinology. J Endocrinol. Quebec City, September 13-15.

9. Thomas G, Frenoy N, Legrain S, Sebag-Lanoe R, Baulieu EE, Debuire B. 1994 Serum dehydroepiandrosterone sulfate levels as an individual marker. J Clin Endocrinol Metab 79:1273-1276.

10. Rotter JI, Wong L, Lifrak ET, Parker LN 1985 A genetic component to the variation of dehydroepiandrosterone sulfate. Metabolism 34:731-736.

11. Corpéchot C, Robel P, Axelson M, Sjövall J, Baulieu EE. 1981 Characterization and measurement of dehydroepiandrosterone sulfate in the rat brain. Proc Natl Acad Sci USA 78:4704-4707.

12. Baulieu EE, Robel P. 1998 Dehydroepiandrosterone (DHEA) and dehydroepiandrosterone sulfate (DHEAS) as neuroactive neurosteroids. Proc Natl Acad Sci USA 95:4089-4091.

13. Vallée M, Mayo W, Darnaudéry M, Corpéchot C, Young J, Koehl M, Le Moal M, Baulieu EE, Robel P, Simon H. 1997 Neurosteroids: deficient cognitive performance in aged rats depends on low pregnenolone sulfate levels in hippocampus. Proc Natl Acad Sci USA 94:14865-14870.

14. Rudman D, Shetty KR, Mattson DE. 1990 Plasma dehydroepiandrosterone sulfate in nursing home men. J Am Geriatri Soc 38:421-427.

15. Berr C, Lafont S, Debuire B, Dartigues JF, Baulieu EE. 1996 Relationships of dehydroepiandrosterone sulfate in the elderly with functional, psychological, and mental status, and short-term mortality: A French community-based study. Proc Natl Acad Sci USA 93:13410-13415.

16. Morales AJ, Nolan JJ, Nelson JC, Yen SSC. 1994 Effects of replacement dose of dehydroepiandrosterone in men and women of advancing age. J Clin Endocrinol Metab 78:1360-1367.

17. Baulieu EE. 1996 Editorial. Dehydroepiandrosterone (DHEA): a fountain of youth? J Clin Endocrinol Metab 81:3147-3151.

3. Industrial Responses to Aging

PHARMACEUTICAL, BIOTECHNOLOGICAL AND NUTRITIONAL INTERVENTIONS IN THE DISORDERS OF LONGEVITY

Hannes B. Stähelin

University Hospital
CH-4031, Basel
Switzerland

INTRODUCTION

The demographic transition observed in developed market economies and increasingly also in emerging countries leads to a rapidly growing proportion of old and very old citizens. In European countries, the absolute number of citizens aged 80 or over has multiplied by a factor of 5 since 1950. Healthy aging and preserving autonomy and quality of life into very old age will be the leading health issue in the 21st century.

There is no doubt that this historically unique development is the result of medical progress, better socioeconomic conditions, better education and public health conditions. Nevertheless, biological age-related changes diminish many necessary adaptive functions and put aging individuals at higher risk for stress. Age-related physiological changes, individual genetic factors and environmental challenges lead with increasing age to chronic diseases.

CONTRIBUTIONS OF THE PHARMACEUTICAL, BIOTECHNICAL AND NUTRITION INDUSTRY.

The changing pattern in diseases observed during this century is clearly related to changing dietary and lifestyle habits (1). Gastric cancer, very common in the first half of this century, is now a relatively rare disease, whereas colonic cancer and, as

Longevity and Quality of Life, Edited by Butler and Jasmin
Kluwer Academic / Plenum Publishers, New York, 2000.

a sequela of the smoking epidemic, lung cancer have become very common. A decreasing incidence and prevalence of coronary heart disease and stroke observed over the past 30 years is not only the result of improved medical treatment, but also of healthier diets (2).

At the center of this health revolution is the molecular biological understanding of the mechanisms of disease. Today, this allows us to develop tailored preventive nutrients. First successful results are the fortification of food with folic acid to decrease neural tube defects, and by lowering the plasma homocystein concentration, the risk of atherosclerosis (3).

A better understanding of the individual risk by identifying particular genetic and environmental risk constellations will be possible in the near future. New types of food will be made available for the prevention of chronic disease. The understanding of how nutrients exert their effects on the cellular and nuclear level will specifically correct metabolic defects in the future. The fields of nutrition science and pharmacology will increasingly converge, resulting in the development of "nutriceuticals," nutrients that may be classified both as foods and pharmaceuticals.

One of the inevitable challenges of old age is decreasing physical strength, as well as visual and hearing impairment. New tissue-compatible materials, information technology, miniaturization and progress in anesthesia and surgery have allowed enormous advances, for example, in orthopedics, cardiology, eye and ear disease. Let us remember that a substantial proportion of impaired functioning in old age is related to impaired vision and hearing (4). These measures will help the aged maintain mobility.

Secondary prevention and treatment of acute disorders in old age will be improved by new drugs. In acute coronary heart disease, and to some extent in stroke, effective treatment is possible, thus preventing subsequent organ failure. Osteoporosis may now be effectively prevented (5). Further progress is imminent.

Increasing longevity puts aging individuals at a particular risk of acquiring neurodegenerative diseases, particularly Alzheimer's disease. The understanding of the pathophysiology of Alzheimer's disease already offers specific treatment options (6). Since surviving cancer and cardiovascular disease inevitably increases the risk of neurodegenerative brain disease, effective strategies are urgently needed.

Some strategic options will be mentioned briefly. Since the true cause of Alzheimer's disease is still unknown, primary prevention is difficult. However, it is well established that good education and empowerment of the individual by health control lowers the risk of developing dementia. The same principles that retard atherosclerosis and the development of cancer also maintain better cognitive function into old age. Thus, diet (and in the future, new types of food) will be important.

One future development may be characterized as support of the impaired brain by outsourcing. Modern life increasingly relies on virtual interfaces in order to control daily life. Instrumental activities of daily life thus become increasingly abstract, and control functions rely on very few cues, e.g., symbols or numbers. Modern life thus puts persons with impaired cognitive functions into increasingly difficult situations. Therefore, information technology has to develop smart systems to help persons with impaired cognition to cope with daily tasks.

Longevity and the longevity-associated health issues will challenge modern societies. They will at the same time create the need for highly imaginative solutions that maintain autonomy and quality of life into very old age.

REFERENCES

1. Willet, WC (1990): Epidemiological Studies of Diet and Cancer, Medical Oncology & Tumor Pharmacotherapy, 7/2/3: 93-97.

2. Braunwald, E (1997): Shattuck lecture - cardiovascular medicine at the turn of the millennium: triumphs, concerns, and opportunities, New England Journal of Medicine, 337/19): 1360-1369.

3. Malinow, MR, Duell, PB, et al (1998): Reduction of plasma homocysteine levels by breakfast cereal fortified with folic acid in patients with coronary heart disease, New England Journal of Medicine, 338/15): 1009-1015.

4. Baltes, PB (1993): The aging mind: potential and limits, Gerontologist 33(5): 580-594.

5. Delmas, PD, and Woolf, AD (1997): Osteoporosis: outstanding issues, Baillieres Clin Rheumatol 11/3): 645-649.

6. McGuffey, EC (1997): Alzheimer's disease: an overview for the pharmacist, Journal of the American Pharmacological Association, Washington, NS37(3): 347-352.

NUTRITIONAL INTERVENTION TO HELP PREVENT AND CURE LONGEVITY DISORDERS

Pierre R. Guesry

Nestlé Research Centre
Vers-Chez-Les Blanc
Lausanne 25
Switzerland

INTRODUCTION

Longevity is rapidly increasing in all industrialized countries, with important socioeconomic consequences, (social security and pension funds expenses, cost of health care, etc.) During the last 50 years, longevity increased by about 20 years and the population above 65 is expected to double by the year 2030 in most industrialized countries. This increase in longevity is due to improvement in hygiene and medical care, but also of nutrition. (Fig. 1)

Increased longevity is naturally accompanied by various ailments. Some physical manifestation of aging are benign and result in mainly aesthetic consequences, such as presbyopia, brown skin spots and wrinkles. Others are more severe, disabling and sometimes life threatening, such as osteoporosis, arthritis, atherosclerosis inducing vascular insufficiency, strokes, myocardial infarction and probably also Alzheimer's disease. The causes of death have markedly changed since 1900, with a decrease of infections and an increase of degenerative disease.

We will focus on the more severe disorders, which can be prevented or partially cured by nutrition, especially since preventive diets and nutrients are generally curative as well. However, prevention is much more efficient than cure, which means that one should start thinking and acting to prevent the disorders while still young. This is difficult and represents a challenge both for public health specialists and for marketing experts of food companies.

Longevity and Quality of Life, Edited by Butler and Jasmin
Kluwer Academic / Plenum Publishers, New York, 2000.

ENERGY RESTRICTION

Most books that deal with the role of nutrition in aging report rat experiments during which caloric intake is restricted by 40 percent from the 6th week of life until death

It is always very dangerous to extrapolate to man from studies carried out on rats living in captivity and in general over-nourished relative to wild rats. Application of the idea that caloric restriction of this type might prolong life in man (2) would mean 40 percent restriction of protein and energy intake from the age of four. The consequence would be severe stunting of growth and, quite probably, a decrease in the learning ability of the child (3).

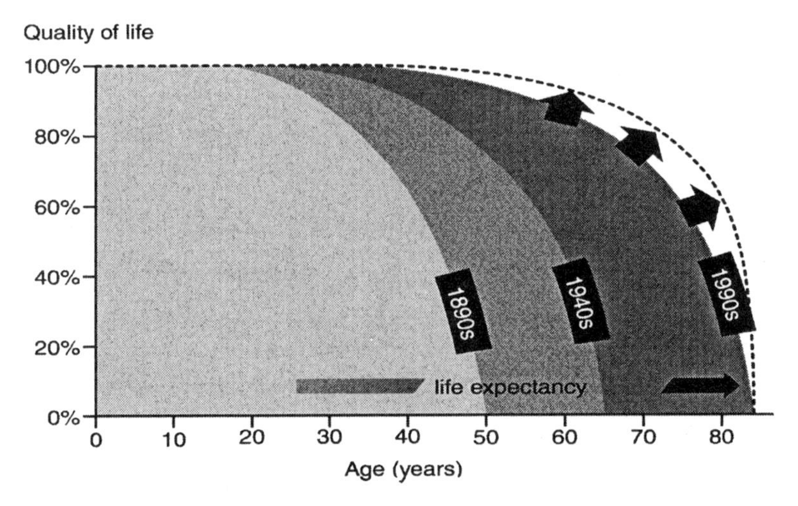

Figure 1. Optimizing quality of life in the later years

The choice would thus be to have a long but abnormal life or a shorter but normal one. Obviously, the situation is not so simple. The dramatic longevity increase observed in Japan since the end of the World War II coincided with a rapid increase in the height of children (4) of about 8 cm at 17 years of age since 1950. This indicates an overall improvement in nutritional status with a general increase of energy/protein intake. (Fig. 2)

In addition, recent studies by Prof. Glickman (University of Victoria, Canada) found no effect of energy restriction on the rate of mutation in rats.

The usual interpretation of these rat studies is, to say the least, overly simplistic, and should only be used to try to better understand relationships of metabolic rate, active oxygen species production coupled with antioxidant defense systems with aging. They should not be seen as models which can be applied to humans. They do indicate, in general terms, that overfeeding precipitates ailments associated with aging and that a moderate energy/protein restriction toward *normal* requirements could be beneficial. (Fig. 3)

Venkatraman (5) showed increased catalase activity, decreased superoxide dismutase activity and no difference in glutathione peroxidase activity in rats' food restricted during exercise.

These studies help us to understand the limits of nutritional modification with regard to age at the beginning of the study, duration, intensity and the potential risks. In fact, recent studies of 64-67 year-olds show that they have a *higher* energy requirement than previously recommended (6).

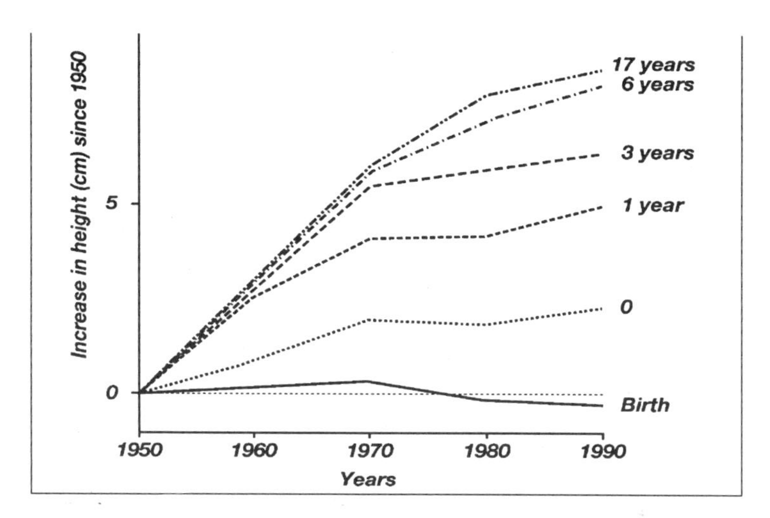

Figure 2. Increase in height of Japanese children during the last 40 years

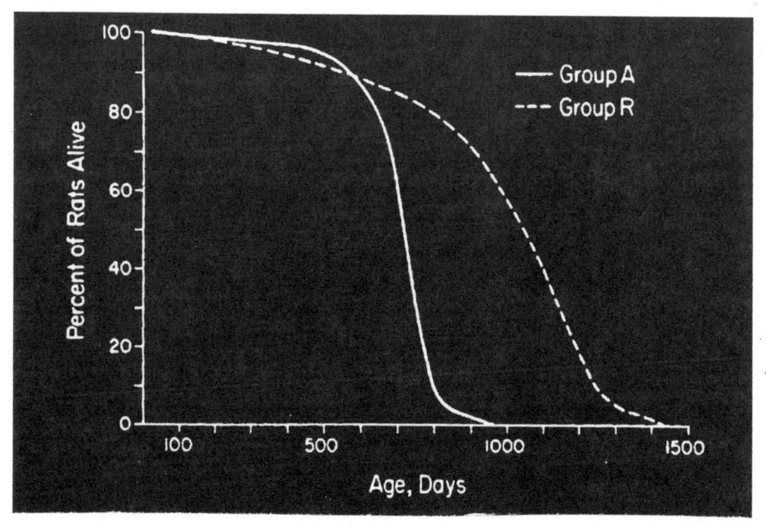

Figure 3. Survival curves of population of male Fischer 344 Rats. Group A (n=115) were fed *ad libitum*. Group R (n=115) were restricted to 60 percent of the *ad libitum* intake from 6 weeks of age on. (J. Gerontol, 1982;32:130-141, modified from BP YU et al.)

PREVENTION OF OSTEOPOROSIS

Prevention of osteoporosis is an excellent example of the benefits of beginning nutritional modification early in life. Preventive therapy probably should start during adolescence, particularly for girls. (Fig. 4)

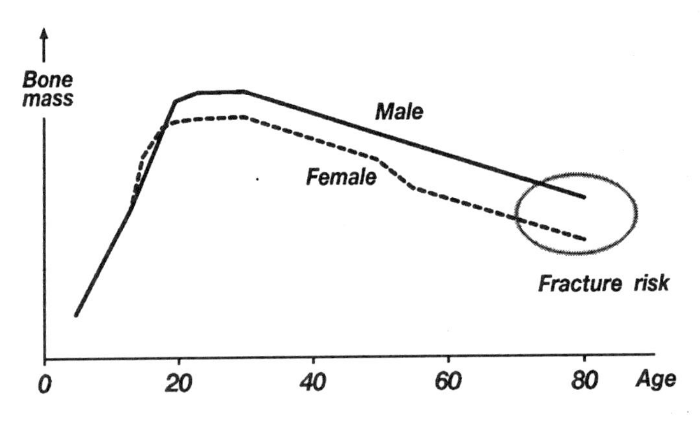

Figure 4. The bone mass is constituted during childhood and adolescence

The studies by Anne-Lise Carrié (7) showed a lasting increase of bone mass when calcium supplementation was given to young girls. The problem is that an adolescent is not sensitized to the risk of osteoporosis and a woman usually starts to be concerned about it when her mother experiences problems or when she herself consults her physician at menopause. Although late, an active supplementation program begun at menopause can prevent the rapid loss of bone mass and delay the moment when bone fragility can create risk of fractures. A recent study (8) shows that an addition of 1.2g of calcium as calcium carbonate per day in women over 60 who have already experienced a spine fracture and previously consumed less than 1g of calcium per day, increased significantly the forearm bone mass ($p<0.001$) and, even more important, reduced by 2.8 times the risk of having a second fracture ($p<0.05$).

Vitamin D supplementation becomes important for calcium absorption and its deposition on the bone matrix because of the decrease of vitamin D skin synthesis by the elderly and intestinal resistance to its action (9).

A recent paper by Trichopoulou (10) shows a good correlation between consumption of mono-unsaturated fatty acid (oleic acid) and bone mineral density which seems promising. Thus, in addition to hormonal treatment of menopause and increased exercise, nutrition plays a very important role in prevention of osteoporosis.

ARTHRITIS PREVENTION

The use of nutrition in the prevention of arthritis is less advanced. Prevention of obesity should bring some benefits but experimental evidence is rare.

Six hundred and forty participants in the Framingham osteoarthritis cohort study (11) underwent knee evaluation by radiography at 8-9 year intervals and the evolution of osteoarthritis was correlated with their intake of different types of vitamins, classified as either antioxidant or non-antioxidant.

There was no effect of either type of vitamin on the appearance of osteoarthritis in knees that were previously normal. However, there was a 3-fold reduction in the risk of osteoarthritis progression for the population with the highest vitamin C intake and to a lesser degree for beta carotene and vitamin E.

What was particularly interesting was that the effect of vitamin C was observed with doses which are still reasonable (around 150 mg/day), only two and a half times the recommended daily allowance.

HIGH BLOOD PRESSURE

High blood pressure has similar effects as aging on kidney. function in rats, inducing protein uria, chronic progressive nephrosis and increased intra-renal vascular resistance (12). The incidence of high blood pressure increases rapidly with age (13), reaching 50 percent after 60 years of age in almost every industrialized country, and coronary heart disease mortality increases very rapidly with high blood pressure. Therapies now exist to efficiently regulate blood pressure with a minimum of side effects.

The role of dietary measures to lower high blood pressure is still the subject of heated debate, particularly the role of sodium chloride restriction for the general population (14). However, the Intersalt Study Group, studying 10,074 men and women aged 20-59, confirmed that an excretion of sodium in urine higher than 100 mmol was associated with a systolic blood pressure higher on average, from 3 to 6 mm of mercury, particularly in the group 40-59 years of age (15). Furthermore, a "modest" salt restriction from 10g to 5g per day reduced systolic blood pressure by about 7 mm Hg in normotensive and hypertensive subjects from 60 to 78 years of age in a randomized double blind trial (16).

In addition, there are many other dietary measures which have proven efficacy on blood pressure. Doubling potassium intake from 50 to 100 mmol per day reduces systolic blood pressure by 3 mm Hg (17). Reducing obesity, particularly abdominal, is highly correlated with a reduction of blood pressure. Reduction of alcoholic consumption will also have an effect.

Most of these dietetic measures can be obtained by reducing animal protein and fat in the diet and compensating by increasing consumption of fresh fruits and vegetables as shown in a recent study (18).

PLASMA CHOLESTEROL

For more than a decade, most scientists have been in agreement that plasma cholesterol and particularly the ratio of LDL cholesterol to HDL cholesterol is a good marker of risk for coronary heart disease in middle aged men (19). However, the discussion is still ongoing regarding the importance of lowering plasma LDL in elderly people (20), mainly because the negative correlation between total serum cholesterol and coronary heart disease mortality disappears at 60 to 70 years of age. Since our objective is to start preventing coronary heart disease early in life, we do not have a problem recommending health and diet measures such as engaging in regular exercise, consuming a diet with less than 30 percent of calories as fat and making sure that these fats are one-third monounsaturated, one-third polyunsaturated and only one-third saturated fat (21), with as little trans fatty acid as possible (22). (Fig. 5)

Among the polyunsaturated fatty acids, priority must be given to n-3 fatty acids (23). Fiber, or at least beta glycans seems to also have a serum cholesterol lowering effect (24), and one should not forget a moderate consumption of wine (25).

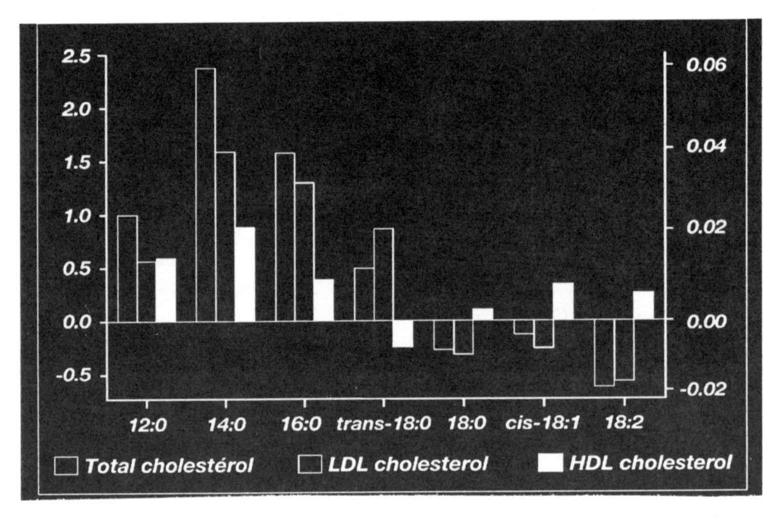

Figure 5. Effects of replacing carbohydrates by fatty acids (1 percent of energy) on plasma cholesterol (PL Zock, MB Katan; Arteriosclerosis and thrombosis, 1994:567-575)

CARDIOVASCULAR RISK

Cardiovascular risk factors are not only high blood pressure and high plasma cholesterol. One should also take into account obesity (particularly abdominal and truncular obesity) and atheroma for which the blood level of homocysteine is an important new nutritional risk factor.

In 1995, Selhub et al. (26) observed an association between plasma homocysteine concentration and extra-cranial carotid artery stenosis. The next step was the demonstration that the plasma level of homocysteine was regulated by folic acid intake (27).

More recently, the possible modulation of cardiovascular risk by dietary folate and vitamin B6 was demonstrated (28) as well as a decrease of thromboembolic accidents following folate therapy (29).

OBESITY

Obesity is the new epidemic in the western world. It began in the US, where up to 50 percent of some segments of the adult population and 25 percent of children (30) are now obese. The epidemic has spread over Europe and to the more affluent sectors of the populations of the less developed countries.

The role of abdominal obesity as a risk factor for atheroma (31) and hypertension is well known, as is its role in non-insulin dependent diabetes mellitus and in certain cancers.

However, obesity has a unique feature as a risk factor for the aging. After 65 years of age (and this age threshold is probably very narrow) increased body mass, both for men and women, decreases mortality in a linear manner up to 85.5 Kg for men and 75.6 Kg for women (32).

Death rates decrease from 46.6/1000 in the group weighing less than 63.9 Kg for men and less than 51.8 Kg for women, to 20.5 per 1000 in the group 74.4 to 85.5 Kg for men and 65.2 to 75.6 Kg for women.

This can probably be explained by the fact that the long-term negative effects of being overweight after 65 years of age become less important than the short-term positive effect of having energy reserves in case of acute diseases which would decrease energy intake. In addition, an older person who is slightly overweight often has a larger choice of food, with more protein and more minerals and vitamins than an older individual who is under weight. Also, an obese individual is less likely to sustain a fracture in the event of a fall because of protective fat cushions.

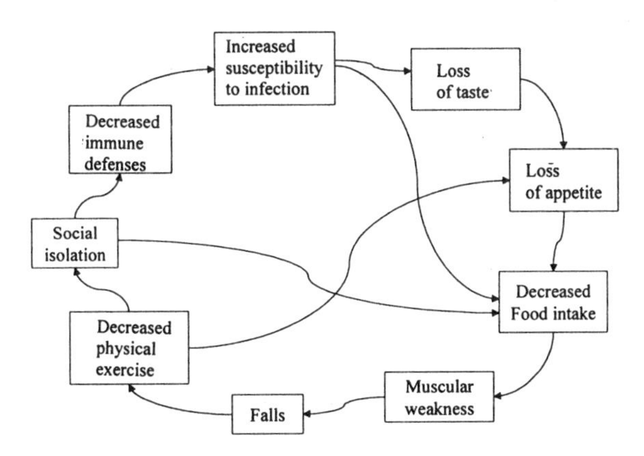

Figure. 6 The vicious cycle of malnutrition in the aging.
(The entry in the circle could be at any point).

In conclusion, obesity should be actively prevented during adolescence and younger adulthood. When the patient is over 65 years, the physician should be less aggressive in prescribing weight loss as long as it does not prevent physical activity. If a slimming diet becomes necessary because of overt articular disease, respiratory problems or hypertension, the diet should contain large quantities of protein and be accompanied by a physical exercise program.

NON-INSULIN DEPENDENT DIABETES

Non-insulin dependent diabetes (NIDDM) constitutes about 85 percent of diabetes in industrialized countries, where the frequency often reaches 10 percent of the population between 30 and 64 years of age (33). This disease, when not optimally treated, will lead to severe eye and kidney diseases.

NIDDM is secondary to cellular resistance to insulin activity which is itself associated with obesity. The best treatment is to reduce the fat mass so as to bring the patient back to an ideal weight while maintaining or even increasing muscular mass.

NIDDM treatment is thus very similar to good control of obesity and its prevention is based on physical exercise and a low-fat diet, rich in cereals and fibers particularly containing beta glycans. These fibers not only promote satiation and slow gastric emptying but, in addition, decrease the rate of glucose absorption after a meal and smooth the curve of hyperglycemia following meals (34). In addition, beta glycans have an hypo-cholesterolemic (35) effect, which can only be beneficial.

DECREASE OF MUSCULAR STRENGTH

The decrease of muscular strength, which happens later with advancing age, is a very important factor, in as much as it can be an entry to the “vicious cycle” of malnutrition of the elderly. (Fig. 6)

Muscular weakness decreases physical activity and thus increases social isolation and loss of appetite, which decreases the amount of food ingested, leading to deficits in proteins, vitamins and trace minerals, which lower the health status.

Of course there are many other entries in this vicious cycle such as bone fracture, infectious diseases, dementia or psychological problems particularly linked to the death of a companion.

Decrease of muscular strength is mainly due to the decrease in physical activity and many authors (36, 37) report very good results in elderly persons who engage in a progressive program of gymnastics and sports.

Decrease in muscular strength is also caused by a decrease in intake of good quality protein. The elderly have been told to decrease their protein intake, particularly animal protein. If they have a small, sometimes functional renal insufficiency, it will decrease further their consumption of animal protein. Moreover, the usual decrease in physical activity and social life of the elderly decreases their appetite and their food consumption.

If there is indeed a decrease in the energy need linked to aging, protein needs remain stable or even increase slightly to about 1 g per kilogram body weight per day. As a consequence, it is often necessary for the elderly to increase food intake and protein consumption, especially when it is expressed in grams per 100 K calories. This increases the need from 1.75 g per 100 K calories in young adults up to about 3 g per 100 K calories for the elderly (38, 39). Physicians should advise older people to select among their normal foods those rich in protein, which are particularly adapted to the taste and needs of the elderly such as yogurt or desserts containing milk and eggs.

CANCER PREVENTION

The World Fund on Cancer Research and the American Institute of Cancer Research on nutrition in cancer prevention have just published a book (40). Our objective is not to summarize the contents of this book but rather to underline that the prevention of cancer through nutrition is neither a utopian dream nor an idea of macrobiotic food fanatics. It has now become a realistic health strategy for everyone.

Intervention studies are now underway in the United States, organized by the American Health Foundation, to try and confirm that low-fat diets rich in vegetables and fruits, can decrease the frequency of recurrence of breast and prostate cancer (41). This has already been demonstrated in mice and rats (42) and is suggested by epidemiological studies in human being (43).

Moreover, the FDA now allows claims that food rich in fiber such as fruits and vegetables can prevent certain cancers and particularly colon cancers.

We know the danger of alcoholism in the genesis of cancer of the mouth, pharynx and liver. Thus, preventive measures include the reduction of alcoholic consumption to a "safe" level of 2 to 3 glasses of wine per day in adult males and 1 to 2 for adult females. Benefits are attained from the antioxidant effect of polyphenols and perhaps also from the positive effect of low levels of alcohol (25).

Natural antioxidants are also contained in fruits (lycopene from tomatoes, vitamin C from fresh fruits) vegetables (beta carotene from yellow and red vegetables), vitamin E from wheat germ, soy bean, eggs and sweet potatoes; phyto estrogen from soya and polyphenols from green tea, coffee and black chocolate or cocoa. In addition, if excessive consumption of polyunsaturated fatty acid increases the risk of breast cancer, by contrast, consumption of oleic acid, mono-unsaturated, as contained in olive oil, seems to decrease this risk (45).

In contrast to regular consumption of fruits and vegetables, attempts to prevent cancer by consumption of drugs containing mixtures of antioxidant vitamins at high dosage have failed (46). This observation leads us to believe that nutritional prevention of cancer is based on a subtle mechanism and a balance of numerous phenomenon linked to synergies between various antioxidants contained in fruits and vegetables. It appears to be a fragile balance, needing all the components to achieve the desired goal.

ALZHEIMER'S DISEASE PREVENTION

Due to the very rapid prolongation of life expectancy and the foreseen doubling of the number of older persons in the near future, Alzheimer's disease, which is already the most common dementia among the elderly (47), is rapidly becoming the most frequent chronic disease in industrialized countries, with severe financial consequences regarding cost and psychological pain for patients and their relatives. Nutritional prevention would thus be of the highest interest.

The chronic soluble aluminum intoxication theory (48) is not fashionable, although some scientific facts, such as the occurrence of typical dementia in young uremic patients receiving aluminum-containing drugs (49) or during accidents of water pollution (50) still support such a hypothesis. With this hypothesis, prevention would be relatively easy by avoiding food and drinks containing excessive quantities of soluble aluminum salt or by making this salt non-soluble, thus less absorbable, by silicating the water.

Today the theory of a vascular etiology is preferred and anti-hypertensive treatment seems to give promising results. Dietetic modification such as the one we discussed, decreasing hypertension and atheroma, as well as antioxidants, could have a long-term prophylactic effect against Alzheimer's disease.

CONCLUSION

This long list of possible mechanisms, some of them still hypothetical and others already proven, shows that we are only beginning to use the various possibilities of nutrition as a means to prevent and cure certain aliments of aging.

Since we are far from knowing all the genetic factors that allow us to identify patients who are at risk, prevention should be proposed to everybody.

We have seen that prophylaxis of certain diseases linked to aging should start very early in life.

Since it is not realistic to expect that the general population will, from infancy onward, consume special types of food, it is the task of the food industry to accept the challenge and improve, progressively and continuously, the nutritional level of food that most people eat, to improve globally the health status of the whole population at all ages. Disease prevention is linked to longevity, thus becoming the positive consequence of general health improvement.

REFERENCES

1. Masoro, E., Food restriction in rodents, an evaluation of its role in the study of aging. Journal of Gerontology, 1988;43:59-64

2. Aspnes, L.E., Kim, M.S., and Weindruch, R. Dietary restriction aging and disease prevention. A discussion of human application. Topics in Clinical Nutrition 1997;12:76-83

3. Scrimshaw, N.S. Malnutrition, brain development, learning and behaviour. Nutrition Research 1998; 18:2; 351-379

4. Takaiski, M., Secular changes in growth of Japanese children. Journal of Pediatric Endocrinology 1994; 7;163-173

5. Venkatraman, J.T. Angkeaw, P. and Fernandes, G., Effects of food restriction on anti oxidant defense system in exercised rats. Nutrition Research 1998; 18; 2:283-298.

6. Roberts, S.B., Energy requirements of older individuals. Europ. Journ. Clin. Nutr. 1996; 50; Suppl 1: S112-S118.

7. Carrié, A.L., et Bonjour, J.P., Osteoporosis as a pediatric problem. Pediatric Clinics of North America 1995; 42; 4: 811-824

8. Recker, R.R., Hinders, S., Davies, K.M., et al. Correcting calcium nutritional deficiency prevents spine fractures in elderly women. Journ. Bone Mineral Research 1996; 11:1961-1966.

9. Dawson-Hughes, B., Harris, S.S., Krall, E.A. et al. Effect of Calcium and Vitamin D supplementation on bone density in men and women 65 years of age or older. N.E.J.M. 1997; 333; 10:670-676.

10. Trichopoulou, A., Giorgiou, E., Bassiakos, Y., et al. Energy intake and mono-unsaturated fat in relation to bone mineral density among women and men in Greece. Prev. Med. 1997; 26 (3):395-400.

11. Mac Alindon, T.E., Yuqing Zhang, P.J., Hannan, M.T., et al. Do antioxidant micronutrients protect against the development and progression of knee osteoarthritis? Arthritis and rheumatism 1996; 39; 4:648-656.

12. Heudes, D., Michel, O., Chevalier, J., et al. Effect of chronic angiotensin 1 converting enzyme inhibition on the aging processes. Am. J. Physiol. 1994; 266: R1038-R1051

13. National high blood pressure education program working group on primary prevention of hypertension. Arch. Int. Med. 1993; 15:186-208

14. Midgley, J.P., Matthew, A.G., Margaret, C., et al. Effect of reduced dietary sodium on blood pressure. JAMA 1996; 275:1590-1597.

15. Elliott, P. Stomler, J., Nichols, R. et al. Intersalt revisited: further analysis of 24 hour sodium excretion and blood pressure within and across population. B.M.J. 1996; 312:1249-1253.

16. Cappricio, F.P. Markandu, N.D., Carney, C. et al. Double blind randomized trial of a modest salt restriction in older peoples. Lancet 1997; 350:850-854.

17. Whelton, P.K. He, J., Cutler, J.A. et al. Effects of oral potassium on blood pressure: meta-analysis of randomized controlled clinical trials. J.A.M.A. 1997; 277:1624-1632.

18. Appel, L.J. T.J. Moore, E. Obarzanek et al. A clinical trial of the effects of dietary patterns on blood pressure. New England Journal of Med. 1997; 336:1117-1124

19. Lipid Research Clinic Program. The lipid research clinics coronary primary prevention trial results: reduction in incidence of coronary heart desease. JAMA 1984; 251: 351-364.

20. Kronmol, R.A. Cain, K.C. Omen, G.S. Total serum cholesterol level and mortality risk as a fuction of age: a report based on the Framingham data. Arch. Int. Med. 1993; 153:1065-1073.

21. Ginsberg, H.N. Barr, S.L., Gilbert, A., et al., Reduction of plasme cholesterol in normal men on an American Heart Association step 1 diet and a step 1 diet with added monounsaturated fat. New England J. Med. 1990; 322 (9):574-579.

22. Willet, W.C., Stampfer, M.J., Manson, J.E., et al. Intake of transfatty acids and risk of coronary heart disease among women. Lancet 1993; 341:581-585.

23. Harris, W.S., Fish oil and plasma lipid and lipoprotein metabolism in humans. Lipid. Research 1989; 30:785-807.

24. Food and Drug Administration: Food labeling: Health claims: Oat and coronary heart desease. Federal Register 1997; 62:3584-3681.

25. Renaud, S. De Lorgeril, M. The French paradox: dietary factors and cigarette smoking related health risks. Ann. N.Y. Acad. Sci. 1993; 686:299-309.

26. Selhub, J. Jacques, P.F. Bostem, A.G. et al. Association between plasma homocysteine concentrations and extra cranial caratid artery stenosis. New Engl, J. Med. 1995;n 332: 286-291

27. Bouskey, C.J. Beresford, S.A.A. Omen, G.S. Omen et al. A quantitative assessment of plasma homocystein as a risk factor for vascular disease. Probable benefits of increasing folic acid intakes. J.A.M.A. 1995; 274:1074-1087.

28. Rimm, E.B. Willet, W.C. Hu, F.B. et al. Folate and Vitamin B6 from diet and supplements in relation to risk of coronary heart disease among women. JAMA 1998; 279:359-364.

29. Candito, M., Bedoucha, P., Jambou, D., et al. Accident vasculaire cerebral avec hyperhomocysteinemie: Arrêt des accidents thrombo emboliques sous folates. Press Med. 1997; 26:1289-1291.

30. Dietz, W., Factor associated with childhood obesity. Nutrition 1991; 7:290-291.

31. Masson, J.E. Colditz, G.A.., Stampfer et al. A prospective study of obesity and risk of coronary heart disease in women. New England Jour. Med. 1990; 322:882-889.

32. Fried, L.P. Kronmal, R.A. Newman, A.B., et al. Risk factors for 5 year mortality in older adults. JAMA 1998; 279: 585-592.

33. Dowse, G.K. and Zimmet, P.Z, The prevalence and incidence of non-insulin dependent diabetes mellitus. In KG Alberti and R. Mazze eds. Frontiers in diabetes research: current trends in non insulin dependent diabetes mellitus. Elsevier Amsterdam 1989; 37-59.

34. Wursch, P. and Pi-Sunyer, F.X. The role of viscous soluble fiber in the metabolic control of diabetes. Diabetes Care 1997; 20:1774-1780.

35. Behall, K.M. Scholfield, D.S. and Hallfrish, J., Effect of beta-glucan level in oat fiber extracts on blood lipids in men and women. J. Am. Coll. Nutr. 1997; 16:46-51.

36. Campbell, W.W, Grim, M.C., Young, V.R., et al. Effects of resistance training and dietary protein intake on protein metabolism in older adults. American Journal of Physiology 1995; 268:E 1143 - E 1153.

37. Nelson, M.E., Fiatarone, M.A., Morganti, C.M., et al. Effects of high intensity strengh training on multiple risk factors for osteoporotic fractures. JAMA 1994; 272:1909-1914.

38. Campbell, N.W. and Evans, W.S., Protein requirements of elderly people. European Journal of Clinical Nutrition 1996; 50:S 180 - S 185.

39. Guesry, P., Dietetic Food in Food Safety: Science, international regulation and control. WHO Ed. Marcel Dekker, New York 1998.

40. Food, nutrition and the prevention of cancer: a global perspective. World Cancer Research Fund / Amercian Institute for Cancer Research Ed. 1997 Washington DC.

41. Wynder, E.L. and Fair, W.R. Prostate cancer - nutrition adjunct therapy. The Journal of Urology 1996; 166:1-2

42. Wang, Y., Corr, J.G. Thaler, H.T., et al. Decrease growth of established human prostate LNCaP tumors in nude mice fed a low fat diet. J. nat. cancer Inst. 1995; 87:1456-1459.

43. Levi, F. Le Vecchia, C., Gulie, C. and Negri, E., Dietary factors and breast cancer risk in Vaud Switzerland. Nutr. Cancer 1993; 19:327-335.

44. Adami, H.O., Mac Laughlin, J.K.., Hsing,, A.W., et al. Alcoholism and cancer risk: a population - based - cohort study. Cancer Causes Control 1992; 3:419-425.

45. Wolk, A., Bergstrom, R., Hunter, D., et al. A prospective study of association of mono unsaturated fat and other types of fat with risk of breast cancer. Archives Internal Med. 1998; 158:41-45.

46. The alpha tocopherol, Beta carotene cancer prevention study group. The effect of vitamin E and beta carotene on the incidence of lung cancer and other cancers in male smokers. NEJM 1994; 330:1029-1035.

47. Stora, F., Alzheimer: l'apparition de nouveaux médicaments marque une étape. Actualité, Innovation, Médecine 1998; 48:14-16.

48. Edwardson, J.A., Alzheimer's disease and brain mineral metabolism. In Nutrition of the Elderly H. Munro and G. Schlierf Ed. Nestlé Nutrition Workshop series vol. 29. Raven Press New York 1992; pp 193-202.

49. Alfrey, A.C., Legendre, G.R., and Kaehny, W.D., The dialysis encephalophaty syndrome. Possible aluminium intoxication. N.E.J.M. 1976; 294:184-188.

50. Neri, L.C. and Hewitt, D. Aluminium, Alzheimer's disease and drinking water. Lancet 1991; 338-390.

GERONTECHNOLOGICAL INTERVENTIONS AND THE HEALTH AND QUALITY OF LIFE OF OLDER PEOPLE

M. Powell Lawton

Philadelphia Geriatric Center
5301 Old York Road
Philadelphia, PA 19141-2996
USA

INTRODUCTION

I speak today to represent the social and psychological point of view on how interventions can benefit the quality of life of older people. Whether the intervention be pharmaceutical, biotechnological, or nutritional, we must define clearly what aspect of the older person's life we intend our intervention to benefit. Despite the emphasis of this entire congress on health, I suggest that treating physical illness and its symptoms constitutes only one part of our task. I shall make a case for examining the benefits of any intervention on aspects of the quality of life that go beyond the domain of health. We should also seek subjective quality of life criteria that indicate not only relief from distress but also enhancement of enjoyment and purpose in life.

HEALTH AND QUALITY OF LIFE

I believe the most active topic in medicine today may be health-related quality of life (HRQOL). The objective presence of illness and sometimes its treatments may cause major changes in the affected individual's life.

Longer life can mean compromised quality if the additional years involve enduring distressing or painful symptoms or if the treatment itself produces such side effects. Useful metrics such as "quality-adjusted life years" (QALYS) have been developed to estimate the value of additional years of life discounted for the conditions that compromise their quality, sometimes referred to as "health utility"

Longevity and Quality of Life, Edited by Butler and Jasmin.
Kluwer Academic / Plenum Publishers, New York, 2000.

conditions that compromise their quality, sometimes referred to as "health utility" (Fanshel & Bush 1970). The recognition of such tradeoffs has been the starting point for consideration of the costs of extended life, expressed in terms of dollar costs of treatment in return for years of life which may be of lowered quality. Inevitably, such accounting leads to ethical questions regarding priorities in treatment, including the consequence that lower life quality may be used as a reason for withholding expensive forms of treatment.

I find three faults with the concept and uses of HRQOL: Quality of life is of only limited concern in the health domain; quality of life is defined by negatives; and the frequent use of quality of life criteria based on the judgments of only healthy younger adults. The first issue is the relationship of quality of life to health. It is difficult to find in the literature a clearly stated rationale as to why the study of quality of life and health requires us to ignore sources capable of enhancing quality of life other than that of health. The most I have seen is the simple statement that there are other domains of quality of life which are not affected by health or are not themselves aspects of health, and are thus excluded by definition from the health-related quality of life discussion.

EMPHASIS ON THE NEGATIVE

The second problem is intimately associated with the first: Why does health-related quality of life research consider primarily negative deviations from a neutral level and rarely positive deviations above the average? The answer seems easy at first glance: Poor health only degrades quality of life. Good health is the norm that we assume and therefore we rarely differentiate health in the average-to-superb range. Health therefore is little involved in a consciously perceived relationship to high levels of quality of life.

The tradeoff principle used in health-related quality of life research allows no input from the positive side. Health-related quality of life research is a tradeoff between more chronological years and fewer quality years. However, I believe of equal importance is the balance between losses as measured by the usual negative HRQOL indicators and the gains in living longer that accrue from non-health-related aspects of everyday life. My assertion is that such gains may offset the distresses consequent to poor health, for some people, some of the time, and in different proportions depending on the person's hierarchy of personal needs and goals. If we measure only health-related quality of life, or worse yet, set medical care policy only on the basis of health-related quality of life, we allow no opportunity for the individual to balance the mix of distresses and pleasures.

These first two major problems with health-related quality of life are thus related: The reason the health-related aspect is too limiting is because it eliminates non-health-related sources of positive life experiences, and therefore the possibility of trading some losses for other gains in assessing one's overall quality of life.

A third criticism of the health-related quality of life stream of research is its assumption that the values attached to states of "good" and "poor" health are the same for all individuals. The weights established for the major measure of quality-adjusted years of life, the Quality of Well-Being Scale, QOWBS, for example, used adults in good health as the standardization group (Kaplan & Bush 1982). How would older people or those in poor health value life under compromised health states? Kaplan has suggested that good and poor health are judged by health status is not a major factor in such judgment. The evidence on this point is scant and mixed. One very compelling finding, however, was that people undergoing kidney dialysis were more tolerant of extended life under reduced quality-of-life conditions than were healthy adults (Churchill et al. 1987). Older persons were also more willing to accept a longer period of hospitalization in return for longer life than were younger people (O'Brien et al. 1977). This study also found that 60 percent of cognitively intact nursing home residents were likely to wish cardiopulmonary resuscitation be applied in case of cardiac arrest and 33 percent wished tube feeding even with permanent brain damage. Unfortunately no data are at hand to allow the direct comparison with the wishes of healthier elders or younger people. Therefore, it still seems worthwhile to explore this question further. Do people's standards for what constitutes an acceptable quality of life show signs of an adaptation process whereby lesser levels become more tolerable as their own health decreases?

Dissatisfaction with these limitations of health-related quality of life has led me to suggest an alternative broader view of quality of life. My definition of quality of life is

"The multidimensional evaluation, by both intra-personal and social-normative criteria, of the person-environment system of an individual" (Lawton 1991).

"Multidimensional" refers to four sectors of quality of life: Behavioral competence, environment, subjective domain-specific quality of life, and generalized psychological well-being. I have depicted it thus:

Behavioral competence represents the social-normative evaluation of the person's functioning in the health, cognitive, time-use, and social domains. This includes signs and symptoms of illness, intellectual impairment, and deficits in activities of daily living, but it also includes satisfying leisure-time pursuits and enjoyable interactions with others.

Environment represents all that lies outside the body of the subject, evaluated also by social-normative standards. The environment may be physical, personal, suprapersonal, or megasocial. Treatment and interventions fall into this sector.

Subjective domain-specific quality of life is the person's own evaluation of the adequacy of everyday life in domains such as housing, community, employment, family, marriage, or leisure activities. Some definitions limit quality of life to this concept. I argue that we need to know *both* objective-normative qualities and subjective qualities before we can understand the total realm of quality of life.

Psychological well-being represents the global level of evaluation of self-in-environment, subsuming much of what is thought of as mental health. In summary, I suggest that quality of life must consist not only of the subjective sectors (domain-specific perceived quality of life and psychological well-being) but also of two sectors which may be assessed independent of the perceiving subject, behavioral competence and the objective environment.

If we plan treatment for one patient, we clearly must know what goals the patient considers important and that patient's judgment about whether a given treatment facilitates the achievement of that subjective goal. But treatment must often be planned for aggregates of people. We can never hope to design a treatment that will be effective in helping all people attain their major goals. We thus make statistical, or probabilistic, judgments about the proportion of people likely to value, or respond well to, a particular treatment. An example from the domain of aesthetic experience is the presence of a symphony orchestra in a city. A proportion of citizens are tone deaf, others don't appreciate concert music. Because we know that a proportion of citizens will be enormously uplifted by hearing a local symphony, the existence of the orchestra constitutes an objective facet of quality of life, with no implication that all listeners will be affected similarly.

These brief descriptions of four sectors of quality of life imply another dimension not fully conveyed by my definition. When the term "evaluation" is used, a polarity between positive evaluation and negative evaluation is assumed: If something is good, it must be not bad, and vice versa. In fact, much of our psychological research in recent decades has made it evident that psychological judgments, and particularly emotional experiences, are evaluated in part independent of their positive and negative components. The first research of this kind, dating from 30 years ago (Bradburn 1969), showed that people could characterize their lives, on the whole, reasonably well in terms of how happy they were. Their emotional states were also measured. Not surprisingly, if they reported more negative emotions, their happiness was less. If they reported more positive emotions, their happiness was greater. The intriguing part of their findings was that negative feelings and positive feelings were not the opposite of one another. In fact, their correlation was zero, in spite of the significant correlation of both with happiness. It is of interest too that increasing evidence from the areas of neuroscience has confirmed the structural and functional bases for the partial independence of positive and negative emotion.

This research is relevant to my topic today because it illustrates how we should be asking not only about the distresses of everyday life in order to determine how much one's remaining life is worth; rather, we should also be asking simultaneously about how much gratification, enjoyment, fun, or elation there is in the person's daily life. I think of this duality as a redefinition of mental health designed to encompass its positive and negative features. Our basic hypothesis is that people engage in a process that has been called "hedonic calculus" (Lawton 1996) whereby all positive and negative input into their lives is transformed into a construct of overall mental health, one of whose facets I have called "valuation of life." In turn, valuation of life, which has both cognitive and emotional components, is the inner motivator for decisions and behaviors that affect selection or rejection of treatments and interventions that prolong or do not prolong life.

Our concept of valuation of life accepts that under many negative mental and physical health conditions, people may feel a diminished attachment to life. At the same time we sought a concept that could also allow the person to account for positive features that might counteract the negative. Valuation of life was the term used that would additionally express elusive concepts such as hope, purpose, sense of future, personal goals, persistence, and meaning to life. Valuation of life is defined as the subjectively experienced worth of the person's present life, weighted by the multitude of positive and negative features whose locus may be either within the person or in the environment. The distinction between positive and negative inputs to valuation of life expresses the frequent clinical observation that to some people, life appears very dear in spite of poor mental or physical health, while to others it appears to be of low value.

Concepts like "purpose in life" (Crumbaugh 1972), "optimism" (Scheier & Carver 1985) and "hope" (Gottschalk 1985) have been developed into scales, but no existing instrument fully represented the important facets noted earlier without also including items that expressed obviously positive or negative mental health. We attempted to purge the items' content of presumptions regarding mental health. How well we succeeded is a matter of degree. Obvious terms like "happy," "satisfied," "discouraged," or "depressed" are absent. There is also no content that refers to life span, death, or health utility. Yet most items can obviously be characterized as representing desirable or undesirable qualities. The extent to which this important outcome is measured independent of its hypothesized antecedents is relative, certainly not absolute. The items composing the Positive Valuation of Life factor are shown in Table 1.

The content reflects our target concepts while carefully avoiding concepts that might be confounded with the *sources* of quality of life or the outcome of health utility expressed in quality-adjusted life years or life-extension wishes. The items do express selected concepts used to denote positive mental health, but they are much more narrowly focused on our core ingredients—hope, purpose, meaning, the future, persistence, and personal goals—than are other measures such as self-esteem, mastery, or optimism.

Table 1. Positive Valuation of Life

Item No.	Item
1.	I feel hopeful right now.
2.	Each new day I have much to look forward to.
3.	My life these days is a useful life.
4.	My life is guided by strong religious or ethical beliefs.
5.	I have a strong will to live right now.
6.	Life has meaning for me.
7.	I feel able to accomplish my life goals.
8.	My personal beliefs allow me to maintain a hopeful attitude.
9.	I intend to make the most of my life.
10.	I can think of many ways to get out of a jam.
11.	I can think of many ways to get the things in life that are most important to me.
12.	Even when others get discouraged, I know I can find a way to solve the problem.
13.	I meet the goals that I set for myself.

At this point the ingredients of my model of quality of life have been defined: There are four sectors, two objective and two subjective, which represent the sources of quality of life. I suggest that the first outcome of the many domains of quality of life is in the realm of psychological well-being or mental health. Valuation of life, which represents the mathematical integral of positive and negative quality of life, is the next more distant outcome. The ultimate outcome is the actual choice regarding prolongation or foreshortening of life, which we suggest is determined most clearly by valuation of life. The sources of quality of life, including health, strongly determine valuation of life, but valuation of life is a processing template that translates the input into wishes and behaviors relating to how long one lives.

I shall briefly describe the empirical data that our research has generated to test these hypotheses. You will see that in the real world many compromises have to be made in turning internal psychological processes and future hypothetical estimates into meaningful measures. For example, because we do not know what our research subjects will do about care and treatment decisions at the end of life, we had to ask them a set of hypothetical questions about how long they might wish to live under various health-compromised conditions—"Years of Desired Life."

Here is what we measured and how we conceived the influences of the various components of our model. The first set of influences are basic background characteristics (age, gender, education, and race). Health is a basic determinant of outcomes. We then chose measures of objective and subjective quality of life to ask what they, in turn, contributed to mental health, valuation of life, and years of desired life. For this purpose we chose three domains of quality of life that have been shown to be important in elders' lives—their activities and interactions with friends and relatives, and the subjectively judged quality of each of those domains. Our earlier research had suggested that these types of engagement with the external world enhanced positive emotion but did not diminish depression (Lawton 1983). Valuation of life was hypothesized to be affected by all of these factors—health, quality of life, and mental health.

Our actual results were quite complex, but they can be summarized as follows:

- Most aspects of quality of life were related to valuation of life.
- However, when we examined the independent effects of all the variables depicted here, there was no independent contribution of either health or depression to valuation of life. Health and objective and subjective aspects of quality of life contributed to valuation of life, but it was primarily because the factors enhanced positive emotions, which in turn increased valuation of life.

Thus our central psychological variable, valuation of life, did behave in most of the hypothesized ways. We can understand a great deal about the person's attachment to life in general through our measure of valuation of life. But what evidence do we have, first, that valuation of life has anything to do with whether people want to prolong or foreshorten their lives; second, what reason do we have to believe that valuation of life is any different from any of our traditional measures of positive mental health?

We have not yet followed these people to learn what health or longevity-related decisions they actually made. The best we could do was to ask them hypothetically about what they might prefer *if* their health declined in particular ways. This is the same principle used in much health-related quality of life research when health utilities are being measured. Table 2 shows the conditions we described to people when we asked, "How long would you like to live if ..." (each conditions were true). Our results were very clear:

- The respondent's present health, quality of life, and mental health were rarely related to the length of time the person would have wished to live under any of the hypothetical compromised health conditions.
- Higher valuation of life and occasionally higher judged quality of present time were related to the wish to live longer independent of all other variables.
- Valuation of life was uniquely related to years of desired life over and above any overlap it may have had with traditional measures of mental health.

Table 2. Years of Desired Life

The question asked is "How long would you like to live if . . ."
(each condition were true)?
No limitations
ADL-dependent at home (no pain, cognitively unlimited)
ADL-dependent in nursing home (no pain, cognitively unlimited)
Confused and ADL-dependent at home (no pain)
Confused and ADL-dependent in nursing home (no pain)
Mild pain (no ADL or cognitive limitations)
Severe and frequent pain (no ADL or cognitive limitations)
Severe pain controlled only by narcotics
Unconscious, no hope of recovery

What do we conclude about assessing the effects of an intervention on quality of life? First, that gerontological research is on the right track when it searches for measurable outcomes in quality of life. Typical criteria include symptoms, side effects, disability in activities of daily living, and depression. These are relevant and should continue to be studied.

Second, other domains, especially those where positive emotion is generated, are relevant to quality of life and to how much the person values life, even if they have no explicit health connection. Positive quality in a non-health domain may counteract a force originating in physical distress that might discount the value of life. We have an interest in the individual's qualitative views of such matters. We have heard people make statements like, "I want to live long enough to see my great-grandchild." "I know that God made this happen and I'm glad to bear it." "Relaxation and meditation don't make the pain go away, but they make it bearable." "I want to spend all the time I can with my wife as long as I can talk reasonably."

Third, our data show that valuation of life is a useful and measurable concept that distills the positive and negative aspects of quality of life and may serve as an indicator of quality of life in both health-related and non-health-related areas of life.

Finally, if we are considering any kind of intervention, whether pharmacological, technological, or nutritional, the clinician must integrate two types of information: First, the research that documents the risks, distresses, and expense of the treatment. Second the clinician must endeavor to learn everything about this individual's circle of family, friends, activities, and the personal goals, weaknesses and religious convictions. What gives the person the strength to endure? Economic models can tell us the broad probabilities of expense and distress, but only a close look at an individual as a unique entity can illuminate the personal value that the he/she places on life.

REFERENCES

1. Bradburn, N. (1969). The structure of psychological well-being. Chicago: Aldine.

2. Churchill, D.M., Torrance, G.W., Taylor, D.W., Barnes, C.C., Ludwin, D., Shimizu, A., & Smith, E.K.M. (1987). Measurement of quality of life in end-stage renal disease: The time trade-off approach. Clinical and Investigative Medicine, 10, 14-20.

3. Crumbaugh, J.C. (1972). Aging and adjustment: The application of logotherapy and the Purpose-In-Life Test. The Gerontologist, 12, 418-420.

4. Fanshel, S., & Bush, J.W. (1970). A health-status index and its applications to health-services outcomes. Operations Research, 18, 1021-1066.

5. Gottschalk, L. (1974). A hope scale applicable to verbal samples. Archives of General Psychiatry, 30, 770-785.

6. Kaplan, R.M. (1982). Human preference measurement for health decisions and the evaluation of long-term care. In R.L. Kane & R.A. Kane (Eds.) Values and long-term care (pp. 157-188). Lexington MA: Lexington Books.

7. Kaplan, R.M., & Bush, J.W. (1982). Health-related quality of life measurement for evaluation research and policy analysis. Health Psychology, 1, 61-80.

8. Lawton, M.P. (1983). Environment and other determinants of well-being in older people. The Gerontologist, 23, 349-357.

9. Lawton, M.P. (1991). A multidimensional view of quality of life. In J.E. Birren, J.E. Lubben, J.C. Rowe, & D.E. Deutchman (Eds.) The concept and measurement of quality of life in the frail elderly. (pp. 3-27). New York: Academic Press.

10. Lawton, M.P. (1996). Quality of life and affect in later life. In C. Magai & S. McFadden (Eds.) Handbook of emotion, adult development and aging. Orlando FL: Academic Press.

11. O'Brien, L.A., Siegert, E.A., Guisse, J.A., Maislin, G., LaPaun, K., Evans, L.K., & Kritki, K. (1997). Tube feeding preferences among nursing home residents. Journal of General Internal Medicine, 12, 304-371.

12. Scheier, M.F., & Carver, C.S. (1985). Optimism, coping, and health: Assessment and implications of generalized outcome expectancies. Health Psychology, 4, 219-247.

IMPORTANCE OF HEALTH IN THE ELDERLY: A CHALLENGE TO BIOMEDICAL RESEARCH & DEVELOPMENT AND TO SOCIETY

D.W. Scholer

Speedel Pharma, Inc.
Petersgraben 35
Ch 4051 Basel
Switzerland

INTRODUCTION

Average life expectancy has significantly increased over the last decades, at least in developed countries, due to improved sanitation, better nutrition, as well as effective prevention and/or treatment of acute diseases. Moreover, control of behavioral health-risk factors, such as smoking, obesity and insufficient physical exercise has recently been shown not only to extend life expectancy further, but also to postpone disability, compressing it into fewer years at the end of life. For a number of chronic diseases affecting particularly the elderly (e.g., systolic arterial hypertension, osteoporosis, Alzheimer's disease), specific treatments were established which reduce serious disease consequences (stroke, heart failure; bone pain, fracture; loss of memory and functional independence).

Additional gain in life expectancy and quality of life can reasonably be expected in the near future, if ongoing scientific progress in our understanding can be translated into innovative diagnosis, treatment/prevention and tangible benefits for patients.

Longevity and Quality of Life, Edited by Butler and Jasmin.
Kluwer Academic / Plenum Publishers, New York, 2000.

GOALS AND STRATEGIES FOR IMPROVING HEALTH IN THE ELDERLY

Healthy aging or an improved quality of life for the aged is a complex goal; it entails:

- attaining a reasonable life expectancy
- ensuring quality of life during the final decades
- postponing and compressing disability into fewer years at the end of life
- maintaining autonomy and social integration.

Aging can be broadly defined as a progressive constriction of each organ system's capacity to maintain homeostasis in the face of challenges, or as a gradual decline in physiological reserve. The therapeutic goal is therefore to maintain function or slow down age-dependent deterioration.

The overall gain in function and quality of life results from a broad range of distinct strategies and their synergistic effects. Contributing strategies are:

- physical and intellectual "mobility" resulting from constant training
- balanced nutrition
- reduced risk factors
- targeted diagnosis and prevention/treatment of aging processes and diseases
- devices/implants/ biosensors
- psychosocial integration and support.

Disease / impaired function	Therapeutic gain	Approach
Parkinson's disease	mobility	drugs (dopamine modulators)
Alzheimer's disease	memory, functional independence	drugs (acetylcholinesterase inhibitors)
Ocular problems	visual functions	drugs + surgery
Hearing impairment	auditive functions	hearing devices
Systolic arterial hypertension	reduced stroke, myocardial infarction, heart failure	drugs (antihypertensives)
Atherosclerosis	coronary functions	drugs (lipid lowering agents) angioplasty, coronary bypass
Arrhythmias	cardiac / brain function	pacemaker
Renal impairment	renal function	drugs (antihypertensives) dialysis, transplantation
Osteoporosis	pain control/mobility reduced fracture	drugs (HRT, calcitonins, calcium, vit. D, bisphosphonates)
Osteoarthrosis	mobility	joint replacement

Figure 1. Recent progress in the treatment of chronic diseases affecting the elderly

Surveying three areas, namely the central nervous system, cardiovascular system and bone and joint diseases, we conclude that substantial and, for the patient, relevant therapeutic gains have been realized. These gains result from different approaches, i.e., drugs, surgical procedures, technical devices and implants. Systolic arterial hypertension and Alzheimer's disease are illustrative of the therapeutic gains achieved.

Isolated systolic hypertension occurs in about 15 percent of persons aged 60 years or more, and was only recently recognized to be of practical relevance.

A large European trial of 4700 patients over 60 years of age was published last year, evaluating the effects of active treatment with antihypertensives versus placebo. This study showed that systolic blood pressure fell in the active treatment group, leading to reductions in fatal and non-fatal stroke by about 40 percent and in fatal and non-fatal myocardial infarction by about 30 percent. Moreover, the study indicates that the treatment of 1000 patients over five years may prevent about 30 strokes and about 50 major cardiovascular accidents.

A second indication which clearly illustrates relevant progress for the patient and his or her family is Alzheimer's disease. The impact of this disease is enormous, affecting about 10 million people worldwide, and it shows an exponential increase with age after 70 years. It is the third most costly disease to society. It is devastating to patient, caregiver and family, bringing with it a slow and gradual worsening of symptoms related to cognition, activities of daily living and behavior. Given these symptoms, the immediate therapeutic goals are to enhance core elements of cognition, improve day-to-day activities, establish a safe and well-tolerated treatment, and reverse or slow disease progression. Indeed, a new generation of so-called acetylcholinesterase inhibitors has been recently introduced, offering Alzheimer's patients a substantial improvement in core elements of cognition and day-to-day activities.

This new class of medicines represents a first, important step toward the treatment of this widespread disease.

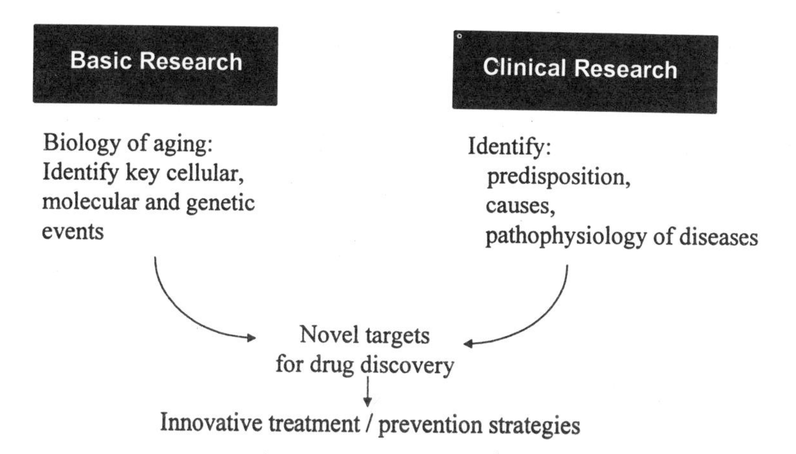

Figure 2. The synergy between basic research and clinical research

RESEARCH APPROACHES/FOCUS ON GENE TECHNOLOGY AND GENETIC RESEARCH

Additional gain in life expectancy and quality of life can be expected in the future from current research approaches and particularly from the combined use of genetic research and gene technology (Figure 2).

Ongoing scientific progress in basic research, particularly the understanding of the biology of aging and identification of key cellular, molecular and genetic events will have a synergistic effect, with progress being made in clinical research, i.e., identification of genetic predisposition, causal factors and the molecular pathophysiology of diseases. Benefiting from this synergy, new targets will emerge for drug discovery which should lead, sooner or later, to innovative treatment and prevention strategies.

Genes are the key focus of today's research. The increasing understanding of the role, function and structure of different genes has far-reaching implications. The access to genetic information has, in fact, revolutionized drug discovery approaches.

Whereas in the past, drugs were primarily identified on the basis of *in vivo* pharmacology or, later, biochemical targets, drug discovery today takes advantage of molecular genetics and molecular biology for target identification, and of transgenic or anti-sense technology for pre-clinical concept validation.

The new perspective of such a rational approach to drug discovery is illustrated in Figure 3.

		Disease		
Therapeutic intervention:	Prevention	Causal	Symptomatic	Palliative
Example: Alzheimer's		Modulation of β–amyloid pathway ?	Alleviate symptoms (memory, activities of daily living)	Basic care nursing home
		Gene therapy ?	Acetylcholinesterase inhibitors	

Figure 3. Disease stages and therapeutic interventions

Disease stages and types of therapeutic intervention are depicted in general terms and shown by way of example for Alzheimer's disease. Palliative treatment (basic care, nursing home) was until recently the only approach available. With the recent introduction of second-generation acetylcholinesterase inhibitors, patients can now benefit from a treatment that alleviates symptoms.

Current research efforts are concentrating on possible approaches at the causal level, particularly modulation of the so-called β-amyloid pathway, which is said to be responsible for the formation of plaques and tangles. For the evaluation of this working hypothesis, the combined use of patient-derived genetic information and of transgenic animal models is crucial.

In fact, some years ago, scientists were successful in identifying in familial early-onset Alzheimer's disease a distinct mutation (the so-called Swedish mutation). Subsequently, scientists at Novartis reproduced this same mutation in a transgenic mouse. In time, these transgenic mice develop morphological changes in the brain which can be visualized by a specific antibody technique. The transgenic Alzheimer's mouse of this particular Swedish mutation develops not only identical brain lesions to those seen in Alzheimer's patients, but also displays functional, learning and memory deficits. It is clear that the availability of genetic information from well-characterized Alzheimer's patients, combined with specific transgenic disease models and modern drug discovery methods, increases the probability of finding innovative therapeutic, or eventually, preventive approaches to this widespread disease.

Genetic research and gene technology, however, provide the key to further progress, not only for Alzheimer's disease, but for many other diseases and indications. The various applications of gene technology are currently at different stages of development/maturity, as shown in Figure 4.

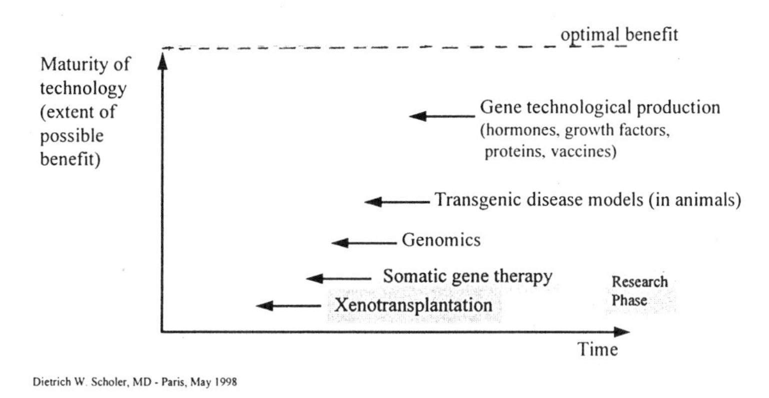

Figure 4. Gene technology: current status and potential

The full benefit of gene technology for the patient is only evident in its application to production techniques which translate into new drugs, like hormones, growth factors or vaccines. Other applications, such as specific transgenic animal models, genomics, somatic gene therapy or xenotransplantation have still to mature before their value to patients becomes tangible.

THE CHALLENGE FOR CLINICAL DEVELOPMENT AND MARKETING

Providing effective and safe treatments for the elderly poses challenges not only for research, but also for clinical development and marketing. These challenges are:

- the move from evidence on efficacy and safety in clinical trials to the "real life" situation, i.e., the issues of selected patients *vs.* heterogeneous patient populations and controlled conditions *vs.* polymedication, as well as the potential for drug interactions and interference with impaired vital function.
- the importance of education/training of family members and health care providers with respect to treatment, dose adjustment and compliance.

Pharmaceutical research and development (R&D) companies approach with confidence the challenge to find new and innovative treatments which improve the lives of the elderly or specifically target the aging process. However, they equally recognize the complexity of this task and are not blinded by any "fantasies of immortality." They operate, not out of an "ivory tower" but in close interaction with universities, and research institutes, with a broad network of R&D alliances (importance as technology platforms or as a specific skill base), and take advantage of the spin-off from broad-based genetic research programs like HUGO. They conduct their R&D in an ethically responsible manner, either internally or together with specific clinical research organizations.

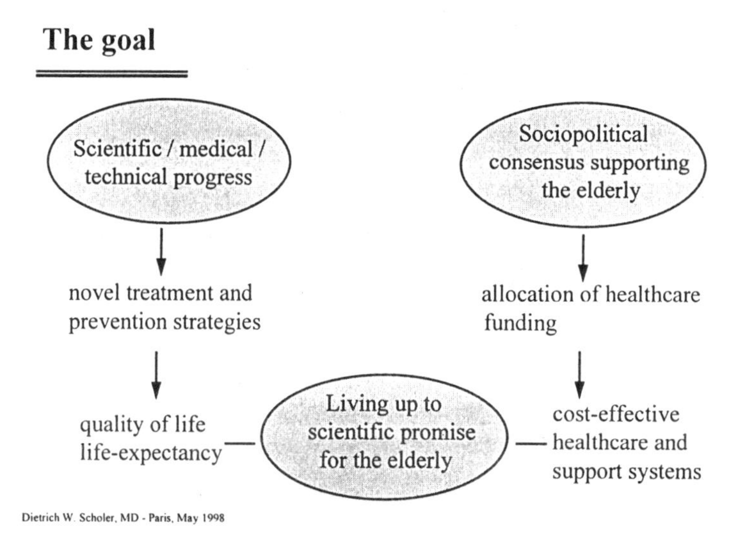

Figure 5. Healthcare costs and sociopolitical challenge

The overall goal is to translate scientific/medical/technological progress into new therapeutics and prevention strategies, thereby improving quality of life and life expectancy, and thus living up to the scientific promise of progress for the elderly. However, scientific progress is only part of the equation. Equally crucial is the sociopolitical consensus, the determination to support the elderly, to allocate healthcare funding and to establish cost-effective healthcare and support systems.

Increasingly, healthcare costs and the economics/politics of longevity have to be taken into account.

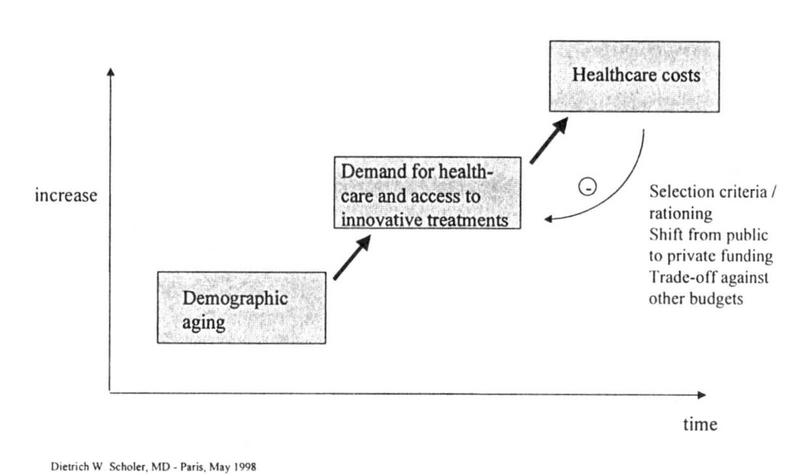

Figure 6. The dilemma: healthcare costs and economics/politics of longevity

Demographic aging leads to an increase in the demand for healthcare and access to innovative treatments, and eventually to an increase in healthcare costs. Cost-containment measures carry certain risks, with a potentially negative impact on the question of who has access to healthcare, e.g., in the form of selection criteria, rationing, demands for a shift from public to private funding, or trade-offs with other national budget categories.

The issue of healthcare costs will therefore form part of the upcoming discussion on how to improve the health of the aging population. Today, about 7 to 14 percent of the gross domestic product of developed nations is spent on health. And drugs represent only a small part, i.e., approximately 10 to 16 percent of total healthcare costs. Indeed, health economic studies show that drugs can be regarded not only as cost-effective, but overall as cost savers.

The fundamental issue is therefore the affordability of improved quality of life. A sociopolitical consensus has to be reached on four relevant questions (as defined by H. Redwood in Pharmapolitics 2000):

- who gets access?
- what quality of treatment?
- what costs?
- who pays?

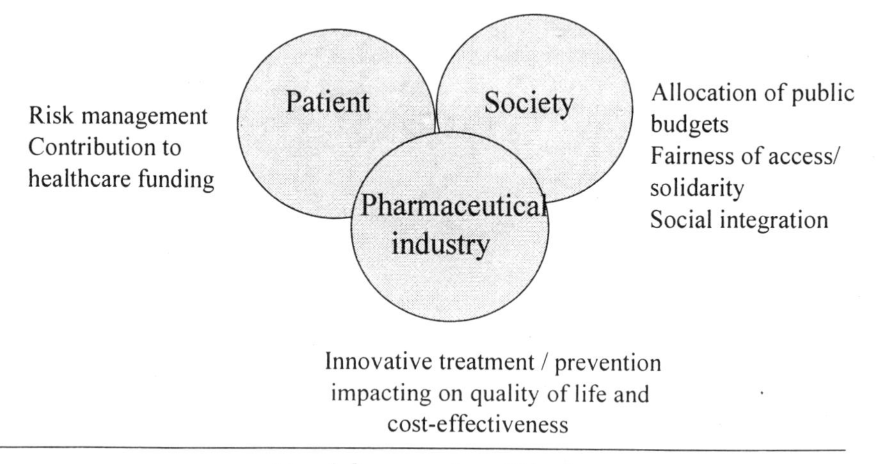

Figure 7. Healthcare for the aging: a growing challenge for society and the pharmaceutical industry

CONCLUSION

Healthcare for the elderly is a growing challenge for society and for biomedical research. The continued use of genetic information from patients, specific transgenic animal models, rational drug design and careful clinical investigation should converge and increasingly advance the effectiveness of symptomatic treatments, but they should also lead to new strategies for causal treatment and prevention, including gene therapy and cell and organ transplants. Progress in biosensor and implant technologies will facilitate the fine tuning or replacement of specific functions.

The well-founded enthusiasm for the fascinating potential of modern biomedical sciences and device technologies, however, should not lead to "fantasies of immortality" or unrealistic expectations. It will be crucial that any feasible translation of a scientific advance into practice for the benefit of patients be measured and judged against a set of criteria such as:

- quality and dignity of life, individual consent, ethical concerns;
- equity and fairness of access;
- cost-effectiveness and economic affordability for the patient and/or society at large.

It is therefore clear that the overall goal of promoting long life and improving quality of life and care, as well as better social integration of the elderly, can only be achieved through a broad sociopolitical consensus and through coordinated efforts on the part of society, national or private healthcare budget providers and innovative research-based pharmaceutical companies.

Quality & Quantity of Life
1. Longevity and Quality of Life

ETHICS AND ENTRANCES

Charles Richard Boddington Joyce

Beim Lindenbaum
Allschwil CH-4123
Switzerland

Without civic morality, communities perish; without personal morality, their survival has no value.

Bertrand Russell (1)

INTRODUCTION

We may not feel as pessimistic as Macbeth that:

Life's but a walking shadow, a poor player
That frets and struts his hour upon the stage
And then is heard no more, (2)

but however long the player's life, he (3) must at some time make his exit from the stage. Every attempt to prolong life is finally bound to fail, but there is no reason not to try to improve or at least maintain its quality until its last days, hours and minutes. The dying individual may even be helped to enter into an entirely new activity, or at least to maintain his dignity. Abiding respect for his wisdom is of the greatest importance when the experience that he acquired during a long life is seen as irrelevant by successors living in a time of increasingly rapid change. As Logan Pearsall Smith puts it: "The denunciation of the young is a necessary part of the hygiene of older people,and greatly assists the circulation of their blood."(4) Or La Rochefoucauld, more appositely in the present case: "Old people like to give good advice, as solace for no longer being able to provide bad examples."(5)

Long before the duration of life began to be considered by other than theological imaginations, it had ceased to bear comparison with that achieved by

Longevity and Quality of Life, Edited by Butler and Jasmin.
Kluwer Academic / Plenum Publishers, New York, 2000.

Methuselah. By such ancient standards, even its later increase during, say, the last two hundred years appears insignificant: the average life still lasts only a tenth of Methuselah's. While the purposes of medical research were largely concerned until recently with the need to prolong life at any cost, the quality of life (than which little, if anything, is more important to patients, especially the aged) (6) was neglected until some thirty years ago. The realization that duration affects quality no less than quality affects duration is even more recent.

This essay attempts to classify available as well as potential methods of prolonging life while maintaining or improving its quality. We shall examine these broadly, in terms of their potential benefits or burdens, and also briefly consider the morality of their application, for discussion of the duration and quality of life must pay attention to their ethical aspects. Quality and quantity are far from equivalent and may even antagonize each other, as individual needs for both may not correspond to those of society, or those of developing nations with those of over-developed countries, and the needs of the present with those of the future. Such conflicts are as often forgotten as distress caused by the use of sophisticated ways to prolong life if its significance is ignored. Beneath all these lies a basic confusion of quality and worth, (7) of individuals as of society as a whole.

We shall draw particular attention to potential conflicts that concern the individual–not only the basic conflict between the desire to live long and live well, but between his state of mind at the time of making a "living will" or "advanced directive" and its state when the time comes for that directive to be obeyed, by which time his opinion may have profoundly altered. Paraphrasing Huxley (who had in his turn distorted Descartes), it may be true for some to say "I am, and therefore wish to God that I were not," (8) but others may feel just the opposite–"I scarcely am, and therefore wish to God I were in better state."

The following description of possible ways of intervention is manifestly personal, and therefore unlikely to satisfy everyone (or perhaps anyone at all). Its crude evaluations of their possible consequences will clearly be even more contentious, and the ethical opinions consequently more so still. The discussion may nevertheless serve to provoke debate.

As space is too limited to consider in relation to each the conflicts to which attention has just been drawn, the patience of the reader is earnestly sought while watching the author try to extricate himself from the moral morass into which he is about to blunder.

PHYSICAL TREATMENTS

Exercise

Physical, psychological and spiritual means can be distinguished from each other. The list begins with an unexceptionable illustration. Physical exercise is certainly ethical (unless the subject suffers from cardiac insufficiency and is smuggled blindfold to the starting line of an Olympic marathon). But this facetious observation does point to the fact that exercise can shorten rather than prolong life. (Even the mere suggestion of exercise, quite apart from its actual performance, may impair the quality of life of subjects like the present author.)

Medical and Surgical Interventions

Medical and surgical interventions upon acute conditions are intended to prolong life, but improvement in its quality is rather the object of intervention in chronic illness. The relief of signs and symptoms, especially pain, is almost always ethical, and if successful will almost always improve quality of life. But its effect upon longevity is not so clear-cut. Prophylactic intervention, on the other hand (such as the preemptive hysterectomy in women at high risk of developing uterine cancer in later life), presents an almost opposing picture. Successful prophylaxis will probably lengthen life, but will sometimes make it less livable. The ethics of a prophylactic intervention, particularly surgical, appear to be more debatable.

Organ harvesting and organ transplantation also present contrasting ethical problems. Harvesting has no chance of lengthening the life of the donor; indeed, for involuntary donors, such as prisoners (including political prisoners) in certain countries, it may reduce the life expectancy of the donor to zero. However, the joy of willingly giving to a partner, a friend or even an unknown beneficiary may contribute a great deal to the quality of life of the benefactor, although anxiety about the consequences (of living with a single kidney, for example) may diminish the pleasure of unselfishness.

The intention of organ transplantation is clearly to lengthen life, with a consequent chance of improving its quality. But the procedure may of course be unsuccessful, as may any other; and even if it is successful, the depression sometimes experienced by patients who receive new kidneys is well recognized. Nor can the ethical question of whether to transplant always be answered with an unequivocal "yes" even if tissue is donated by an animal–perhaps not even if the substituent is artificial.

Genetics

In the opinion of Steve Jones, "Genetics is, in the public mind, rather like Christianity. It is a matter of faith; a curse or a salvation, promising or threatening its believers according to taste. In fact, genetics has achieved little and told us almost nothing about human affairs that we did not know before. It... has given rise to many works of exegesis offering eternal life in molecular paradise or (choose your church) everlasting damnation for those who follow the broad path down a double helix to Hell." (9)

Artificial insemination presumably lies somewhere between medical and genetic methods, and consideration of creation and use of gamete banks should perhaps also be differently approached from the point of view of donors and recipients. There is unlikely to be a direct effect upon the longevity of the donor, although it must be assumed that the effect upon his quality of life is always positive, as it is likely to be for the recipient (at least until the outcome reaches adolescence).

Recent developments have made it possible for postmenopausal women to bear children. The expectation, presumably, is that the mother's duration and quality of life will be increased, but the outcome will not always be the one that was expected. In any case, the ethical implications are obscure, or likely to be negative from the point of view of the child with one or more overage parents.

Other methods–genome manipulation, cloning and telomere extension, for example–are not yet out of the experimental stage, and it would therefore be premature to predict their consequences. However, it is legitimate to demand even more strongly than in the previous examples, or for the psychological methods that will be discussed next, that much more thought is needed about the wisdom of prolonging life as well as improving its quality by genetic means.

PSYCHOLOGICAL METHODS

Exercise

Psychological exercise includes such educational methods as memory training, practice in decision-making and the solution of judgmental problems, the maintenance of existing creative behaviors or the inculcation of new ones, and the drawing-out of previously unsuspected talents. This may not prolong life, but is likely to improve its quality unless a lack of success induces disappointment or depression. There can hardly be any objection on ethical grounds.

Interaction and Work

The maintenance, or increase, of social interaction seems a thoroughly ethical way of affecting quality of life, but the effects upon longevity are more difficult to predict. Frail physique or diminishing psychological strength and ability to cope may shorten life if greater demands are made, but the setting of new tasks and targets may well work beneficially in an individual who is still relatively strong in body and spirit. The importance of companionship to the elderly, even the possibility of a new love, cannot be overemphasized, but whereas companionship may survive a wide difference in ages, it is unlikely that sexual attraction will be able to do so for very long.

Withdrawal

This is the very opposite of interaction. It is a deliberate or unconscious reaction to stresses of all kinds, real or imagined, and it can hardly be expected to prolong life, although its effects upon life quality may, at least for a time, be beneficial. It is of interest that divorce statistics among the elderly, like those in younger age groups are also rising; at least in Switzerland they are now said to be rising faster among the elderly. A decision to withdraw may be unethical, in so far as it has an adverse effect upon others, and it may be ethical for another to intervene in such a way as to prevent it, although this is not inevitably so. The distinguished existential analyst Heinz Lehmann is known on at least one occasion to have agreed with his patient that suicide was the only possible solution to her problems. Such withdrawals must be carefully distinguished from the spiritual paths, to which this classification leads next.

SPIRITUAL METHODS

Retreat and Meditation

The effects of retreat and meditation are much more likely to be positive. Although it is not possible to be sure if the practices have any effect upon longevity, they are more likely than not to improve quality. They seem to be unexceptionable, too, from the ethical point of view, unless imposed upon rather than sought by the individual.

Return to the Faith; Conversion; Rebirthing

Older people, especially perhaps women, may return to their church when the loneliness of later life imposes upon them. The likely effects upon their quality of life are interesting but too complex to discuss here. The belief in eternal damnation, for example, is apparently not yet extinct. It may lengthen life by inspiring a desire for time to repent real or imagined sins–or for delaying confrontation of their consequences. Conversion and rebirthing have something in common, although in the usual sense conversion implies changing a system of belief whereas rebirthing implies acquiring access to a more satisfactory life by "reliving the moment of one's first breath and releasing the trauma of it." (10) It is impossible to be dogmatic about the effects of either upon longevity or quality. The ethical position of both, too, is far from clear. If the necessary processes are undertaken for the benefit of the subject, well and good, but if they are undertaken because of outside pressure from the healer, the religion or the sect itself (as in the case of Scientology among others), then they are clearly unethical.

Reincarnation

It may surprise some people to see that reincarnation is included as a possible means of prolonging and improving life. There are more ways than one of reincarnation–all but one of which have been known, and practiced, for hundreds if not thousands of years. The oldest of them is procreation, and the second oldest the perpetuation of at least a part of the individual personality by leaving influences, whether by art or science, for the future to treat sometimes more tenderly than their progenitors. Some of us try to preserve our own quality of life at the expense of that of our audiences by continuing to do whatever it is that we do for far too long.

Genetic science may contribute a totally different kind of reincarnation, in the form of cloning. As space becomes more and more precious, preserved samples of individual DNA may replace cemeteries as well as the genealogies that were once found in the family bible, by providing to our descendants a family tree the branches and twigs of which it may become possible one day to bring back to life, at least in terms of anatomy and physiology though not of experience. Long-term sperm banks are already available, if that is the correct word, but ova can at present be kept only for a more limited period. Some ethical problems have been briefly mentioned already, but the difficulties inherent in obtaining permission to make use of such deposits, particularly when those concerned and their relatives have been lost from contact for whatever reason should also be noted. On the other hand, the possible use of banked fetuses is already being investigated. Though it poses ethical problems that are perhaps even more severe, these have little to do with the subject under discussion: quality of later life.

At the back of all these speculations lies the concept of reincarnation *proprement dit*. There is no doubt that, although it is not susceptible to scientific demonstration, still less to human desire, control or even belief, reincarnation undoubtedly represents, in a sense, an extension of living, although whether such perpetuated lives may bear improved or diminished quality few of us have any certain means of knowing. The possibility finds its place in this list, in part for the sake of comprehensiveness, but also because there seems to be at least one ground–it can scarcely be called a reason–for belief in the real possibility. Would it not be pure arrogance, recklessness, to ignore those wise ancients who devoted more time to these matters than we have done since then? Surely reincarnation may be said to prolong longevity, in a certain sense; though the quality of the lives that it may thus perpetuate is hard to predict. Eastern ideas of reincarnation, especially the Buddhist concept of what may be called "modular" reincarnation must be mentioned here, however briefly, for they make possible the acquisition of further credit towards the attainment of Nirvana by pilgrimage and valuable deeds. The concept of reincarnation apparently has no place in Islam.

Comment upon the ethics of reincarnation seems irrelevant, inappropriate and, indeed, impossible. (11)

CONCLUSION

Some of us who are involved in research of this kind do not always consider carefully enough the ethical problems it poses. Our political masters, too, should be considering more than the economic consequences of world population bursting beyond its limits. There are too few signs that they are doing so. Real progress must come from individual concern with the significance, the meaning of life, and indirectly its quality. Otherwise, the prolongation of life is not a desirable end in itself. Attempts to improve quality are not likely to meet opposition, especially from those whose expectations of quality or length are slight. But agreeing upon, let alone achieving, the means to this end is not easy. Communities as well as nations need to define their purposes more clearly to make discussion possible.

There can seldom be an ethical objection to successful methods of improving quality of life by means that are at present available. But intensive research upon the use of spare parts (corneas, hearts, bone marrow, etc.) and pharmaceutical lubricants to regrow hair (Propecia), or restore "positive thinking" (Prozac), (12) and especially potency (Viagra) and morning-after contraception, strongly resembles abolishing one traffic bottleneck by road-widening only to create another one further down the road. The Vatican condemns chemical contraception but not potency stimulation by Viagra: yet another demonstration that we have so far learnt little about the pharmacology of rational thinking. Definitions by individuals or groups that are not religious in one sense or another have seldom been explicit; their implicit sense can seldom be separated from a voracious appetite for power. Do such phenomena as the growth of the Cafés des Philosophes promise to renew the essential endowment of the group with the wisdom of individuals?

As Kant has it: "Morality is not really the doctrine of how to make ourselves happy, but how we are to be *worthy* of happiness."(13) Recently, Richard Jenkyns has seen the dilemma as "the threat that we shall acquire godlike powers over the making and shaping of human beings which we lack the moral preparation to use wisely (if indeed we should use them at all)." (14) Or, as Bertrand Russell concisely summarized it, fifty years ago: "Without civic morality, communities perish; without personal morality, their survival has no value."

ACKNOWLEDGMENTS

This essay has greatly benefited from discussions with Dr. Ismene Kostaki-Joyce.

REFERENCES

1. Russell, B., Lectures on Ethics. University of Cambridge, 1947/1948.

2. Shakespeare W. Macbeth. London:1606;V:v, 24-26.

3. Strunk, W. and White, E.B., The Elements of Style. 3rd ed. New York: Macmillan, 1979. p. 60.

4. Pearsall, Smith, L. Afterthoughts, 1931 (Merriam-Webster Dictionary of Quotations, 1993).

5. De la Rochefoucauld F. Réflections ou Sentences et Maximes Morales. Paris, 1665.

6. Joyce, C.R.B., Health Status or Quality of Life: which matters to the patient? Journal of Cardiovascular Pharmacololgy, 1994; 23, S26-S33.

7. Shaw, G.B., The Doctor's Dilemma, 1946; London: Penguin.

8. Huxley, A. Brave New World, 1932; London: Chatto & Windus p. 273.

9. Jones, S. In the genetic toyshop. New York Review of Books,1998;45,7:14-16.

10. Orr, I. and Roy, S. Rebirthing in the New Age. Milbrae, CA: Celestial Arts, 1977.

11. Joyce, C.R.B. and Stewart, T. R., Applied research on judgment. Acta Psychol,1994;87:217-227.

12. International Herald Tribune. On the Pharmaceutical Frontier. April 27, 1998; pp 1 & 6.

13. Kant, I. Critique of Practical Reason. 1788.

14. Jenkyns R. 2001. New York Review of Books,1998;45, 9:4-7.

DIGNITY, WELL-BEING AND QUALITY OF LIFE

Jonathan M. Mann*

Implicit in the World Health Organization's definition of health as "physical, mental and social well-being" is a connection among the concepts of dignity, well-being and quality of life. Unfortunately, long after this definition was promulgated, the ability to describe physical, mental, and particularly social well-being in meaningful terms remains elusive. And since dignity is not actively discussed, it can neither be recognized as a health-defining condition, nor as a condition fundamental to quality of life.

What is dignity? Having failed to find a broadly accepted definition, a series of investigations were undertaken. The following discussion is derived from observations and ideas generated through talks with people from a variety of backgrounds and cultures. The proposals which follow are intended only to be initial explorations into a profoundly important "terra incognita" of health.

Dignity seems to flow from two components: internal ("how I see myself") and external ("how others see me"). The common denominator is being seen, and the nature, or quality of this perception. While a few—Nelson Mandela comes to mind—seem able to sustain personal dignity despite powerful external negative views, and while some seem to lack personal dignity despite strongly positive external perceptions, for most people, a sense of personal dignity reflects a dynamic process—often under strain and flexible adjustment—involving both internal and external perceptions.

There is abundant evidence to suggest that dignity is of vital importance to the individual and society. People and societies spend considerable energy on the daily effort to protect, enhance and sustain their dignity. To illustrate: in organizations, minor perceived slights easily escalate into severe interpersonal problems; questions of precedence on lists, in seating arrangements, and in distribution of material and symbolic rewards are endowed with enormous significance; "saving face" is a widely felt concern which gives rise to complex societal patterns of behavior; and the verb "to diss," meaning "to disrespect" is a sufficiently recognizable and commonplace phenomenon in American life to warrant its entry into mainstream language.

The meaning of dignity for health and quality of life can be approached indirectly, by focusing upon violations of dignity. This is a rich source of deeply held beliefs and feelings. From many discussions of personal experience with violations of dignity, a provisional taxonomy of dignity violations can be proposed, and a sense can thereby be obtained about the impact of dignity violations on well-being.

Longevity and Quality of Life, Edited by Butler and Jasmin.
Kluwer Academic / Plenum Publishers, New York, 2000.

One form of dignity violation involves "not being seen," as occurs when a person is not recognized or acknowledged. For example, if we seek assistance from a doctor, or a teacher, and that person does not signal awareness of our existence (by ignoring us, or—in our societies—by refusing to make eye contact or shake hands), dignity is at risk. At the extreme, guards in prisons and concentration camps have allegedly been instructed to look at inmates only in the center of their forehead, and never in their eyes. "Not being seen" seems to be a common form of dignity violation.

A second form of dignity violation involves being seen, but only as a member of a defined group. For example, many women cite situations in which they were told they should, or should not, do something because they are women, irrespective of their individual skills, capacities or interests. Thus, to be told, "as a woman, you should not enter this or that career" denies individuality in favor of a group categorization, and such characterization usually has pejorative implications. Yet even if the group classification is a source of pride—as an African-American, a Frenchman, a Catholic, an Englishwoman—the dignity-injuring element remains, because individual character is denied by subsuming all into a group identification.

A third injury to dignity results from violating personal space. Each culture defines an invisible space around the self (and parts of the body) which, if entered without permission, injures dignity. A slap to the face violates dignity, with its impact on well-being far beyond the usually trivial physical injury; rape is an extreme example of this kind of violation. Personal space also includes certain living and work areas; a sense of dignity violation commonly arises on discovering that personal mail has been intercepted and read. However, personal space can also be entered, with permission, without any loss of dignity, as in loving physical intimacy or in professional medical or psychiatric care.

Humiliation is a fourth form of dignity violation. Humiliation seems to involve being distinguished and separated from the group, or from a social norm. To be "singled out" for criticism (particularly in front of the group), or to have one's singularity emphasized (dressing either too formally or too informally for the occasion; applauding at the "wrong" moment during a concert) may evoke humiliation.

With these broad categories of dignity violation in mind, an exercise has been developed. In a group setting, people are asked to recall a situation in which they felt their personal dignity was violated. It is evident that everyone can easily remember such occasions. After brief reflection, participants are asked to describe what happened and then to speak about how they felt when the event occurred, and how they feel about it now.

Strong, sometimes overpowering emotional responses are evoked and described, including shame, anger/rage, powerlessness, frustration, sadness, disgust, a feeling of uncleanliness, and hopelessness. Not only are these feelings vividly recalled, but simply remembering the event restimulates the emotions attached to it, so that participants describe actively reliving the actual event. Injuries to dignity which occurred decades earlier continue to evoke powerful emotions.

At this point, participants are reminded that they generally live in and were raised in a dignity-supportive environment, in which they experienced occasional lapses of dignity as significant, memorable life events. How then might the quality and duration of impact differ for people who live in an environment characterized by severe, sustained, institutionalized and repetitive violations, who may experience only occasional dignity-affirming situations?

Dignity violations and their significance for health and quality of life may also vary with time—reflecting both personal and societal history. For example, adolescence seems a period of substantial vulnerability to severe impacts of dignity violation. Yet older people are also quite vulnerable to dignity violations. This individual life-cycle vulnerability is likely co-mingled with social history: certain historical periods of heightened or reduced group consciousness or pride may mediate the personal experience of potential violations of dignity.

At present, it is difficult to proceed beyond an intuitive linkage of dignity, health and quality of life. In this regard, an awareness of the importance of dignity for well-being—physical, mental and social—is both evident and veiled. This pre-articulate phase of understanding is actually not uncommon in public health. Consider for example the history of child abuse or domestic violence. In each case, the nature or extent of the problem somehow eluded conscious awareness until a specific process of discovery occurred.

Such discovery seems to involve several steps. First, the phenomenon must be named. When the "battered child syndrome" was reported in the medical literature, the veil of silence started to lift from this enormous global problem. Even more recently, in the context of increased attention to women's health and to violence, domestic violence has started to attract the public health attention it so urgently needs.

Naming makes descriptive epidemiology possible. At this stage, and with necessary attention to definitions and categorization, the issue of concern is often found to be larger and more widespread than was assumed. And precisely at this point —often highlighted by political and media attention—the lack of prior attention to the problem—child abuse, domestic violence—seems hard to understand.

Regarding dignity, health and quality of life, we are now at the frontier—the beginning of the conscious phase of public health attention. For we are realizing that violations of dignity are pervasive events, with potentially severe and sustained negative effects on physical, mental and social well-being. It is evident that much remains to be discovered in the world of human suffering. Just as in the microbial world, in which new discoveries (Ebola virus, Hantavirus, Toxic Shock Syndrome, Legionnaires's Disease, AIDS) have become the norm, so explorers are needed in the larger world of human suffering and well-being. Injuries to individual and collective dignity may represent a hitherto unrecognized pathogenic force of destructive capacity toward physical, mental and social well-being at least equal to that of viruses or bacteria.

Naming, describing, and classifying dignity violations are necessary steps toward this informed awareness. Yet in seeking to identify, measure and understand dignity and well-being, it will be important to be creative. The health consequences of dignity violations are not likely to be expressed through a single biomedically-mediated pathway; dignity violations may not necessarily manifest themselves as a single disease such as hypertension or asthma. The concept of a single agent causing a specific disease, so useful in an earlier era of microbial discovery, does not seem useful here. A more ecologic, AIDS-based analogy seems appropriate, in which dignity violations may reduce resistance or the capacity to respond adaptively to a wide range of environmental stresses so that its effects on well-being may be multiform and diverse and therefore require substantial attentiveness in order to be detected.

In this regard, it is important to recall a remarkable phenomenon in medical care. In most studies, from one-fourth to one-half of the people who visit a general medical practitioner depart without a specific medical diagnosis. During my own medical training I recall that we were subtly socialized to view these people as "wasting our time" or "misusers of the health system," for after all, they had "nothing wrong with them!" However, rather than stigmatizing people for seeking medical care for forms of suffering medicine cannot see because they do not fit into the existing biomedical lexicon, it will be important to acknowledge that the health vocabulary and categories are themselves too narrow to recognize certain forms of suffering.

Future health professionals may look back at the current, limited and narrow understanding of health, and wonder how we could have missed violations of dignity as sources of injury to well-being, and the therapeutic avenues opened by considering methods of strengthening dignity.

How then to proceed, while awaiting research to clarify interactions between dignity and quality of life—while awaiting the more refined study of dignity and health? One approach can be based on the modern human rights movement, and in particular on the first article in the Universal Declaration of Human Rights, which states: "all human beings are born free and equal in dignity and rights." This is the essence of the human rights movement, for "dignity" and "human rights" are seen as inextricably linked. Promotion and protection of human rights in the absence of respect for dignity, is inconceivable. In the end, human rights, like dignity, become meaningful only when people accord to others the rights and dignity they assume and wish for themselves. This concept does not minimize, let alone preclude, important differences between people in terms of skills, intelligence and other capabilities. Yet it establishes a common ground, a set of rights, and a respect for dignity, in which all are equal.

When we reflect on dignity we immediately realize how capable we are of promoting the dignity of others. Once we are consciously aware, we become responsible, within our own sphere, for acting in concrete and specific ways to avoid the subtle and often pervasive acts that deny people their individual identity, invade their personal space, or participate—even passively—in their humiliation. Those in the health professions can identify in the ways services are organized, or care and aftercare are provided, many pathways through which the dignity of others may be injured. This surely must be related to the shocking level of frustration and anger people often express after seeking "medical attention."

Therefore, the concept of dignity and the importance of working to promote and protect human rights should be integral to discussions about quality of life. Promotion and protection of health and promotion and protection of human rights are now understood to be inextricably connected. We must pay attention to this linkage, for it can also provide us with guidance as we seek to promote the quality of life for older people. We can ensure that to traditional methods of risk factor identification, education and services to reduce individual risk, we add an understanding of the societal basis of risk and its reduction.

Thus, the modern human rights movement gives us a powerful conceptual framework of dignity and rights–not older people's dignity and rights, but rather the rights and dignity of older people; our, and their human rights and human dignity. In turn, this will help us as we seek to define, at a personal and societal level, what it means to be human, to be treated as human, and to act human. It is the linkage we can establish between individual and societal destiny.

* Dr. Jonathan Mann and his wife, Dr. Mary Lou Clements-Mann died on September 2, 1998, when their plane crashed off the coast of Nova Scotia. Dr. Mann was a pioneer in the campaign against AIDS and a world renowned leader in the movement to link health issues with those of human rights. At the time of his death, Dr. Mann was the founder and first dean of the Health Sciences School of Public Health at Allegheny University (since renamed MCP Hahnemann University School of Public Health).

HEALTH, LONGEVITY AND QUALITY OF LIFE

François Régnier

Synthelabo Groupe
22 avenue de Galilee
Le Plessis Robinso
N. Cedex 92355
France

INTRODUCTION

The definition of life is sought in abstract considerations. I like the general definition: Life is the group of functions that resist death. This is from the 18th century text by Xavier Bichat: *Recherches physiologiques sur la vie et la mort.*

In 1998, life expectancy is almost three times that found at the beginning of the Enlightenment. However, the quantity of life from which we hope to benefit involves a new consideration, that of the quality of life. Bichat, an anatomist and embryologist who gave his name to a major Paris hospital, was far from predicting such a change in perspective.

During 20th century, industrial society first tried to increase productivity, that is, the quantity of objects produced with an economy of production factors, such as labor and the cost of energy. At the end of this century, the two technologies of the new industrial civilization are those of information and of quality. We now know that although increases in productivity are still of interest, they are complicated by a complementary goal, that of quality or even the total quality of products. Why should the quality of life differ from that which we want to measure in the technical field?

A BACKDROP FOR HEALTH, LONGEVITY AND QUALITY OF LIFE

Just as it is difficult to consider medical preventive measures without social preventive measures, it seems difficult to consider questions of health, longevity and quality of life without consulting the population indicators. Although longevity is higher in the industrial countries than in the developing countries, there are major differences between the industrialized nations. Without overloading this paper with

Longevity and Quality of Life, Edited by Butler and Jasmin.
Kluwer Academic / Plenum Publishers, New York, 2000.

figures (1), it is necessary to put into perspective the population data that throw light on our topic. Here are three examples:

First Sweden, with 17.5 percent of the population over 65 may seem to be the country with the highest aging population. However, it should be noted that the rate of reproduction is 2.1 children per woman in Sweden. This exactly corresponds to the rate of population renewal. In addition, Sweden has a very high life expectancy: 81.2 years for women and 75.5 years for men.

France, with 14.6 percent of the population over 65, has a lower proportion than Sweden. However, as opposed to Sweden, the French rate of reproduction is only 1.7 children per woman. In other words, it is lower than the renewal of the population. Even though the life expectancy is high in France (80.9 for women and 73.1 for men), the overall data taking into account both longevity and population renewal seem less favorable than in Sweden.

The objective references to population data are a backdrop for a study of health, longevity and quality of life. The figures provided for Sweden and France are from 1993. By projection, the West is aging. Jacques Attali (2) provides a third example. In the United States, while only 4 percent of the population was over 65 years old in 1900, this group accounts for 13 percent today and will account for 30 percent in 2025. Seven million Americans will be over 85 years old. This elderly majority will be faced with a potential revolt from the young minorities. From the perspective of society, longevity and quality of life may appear to be conflicting.

FIRST IT'S NECESSARY TO CONSIDER GOODS AS GOODS AND THE ELDERLY AS PEOPLE

When the World Health Organization (WHO) was founded in 1946, it already led the way with its positive definition of health as a state of full physical, mental and social well being and not simply the absence of disease or infirmity. Lisbeth Sachs (3) recommends changing this definition: "Health is a state of full physical, mental and social well being, with or without disease or infirmity." This corrected definition globally takes into account asymptomatic, invisible disease. Feeling in good health and having a high quality of life may also be found in people who suffer from chronic diseases and disabilities. We recommend adding: "different chronic diseases and disabilities whether or not related to longevity."

The French civil code, borrowing from Roman law, makes a basic distinction between people and goods. Equating goods with people now seem to be a magic or supernatural frame of reference that no longer has a place in modern medicine. Today, medicine relies on the experimental demonstration of proof and the dynamic current of evidence-based medicine. Considering goods as goods and people as goods would be a technocratic error or even totalitarian despotism. In fact, Lenin and Stalin wrote that man is a cog in the machine to build communism.

The main challenge of medicine today is that goods have to be considered as goods and the elderly as people, without confusing the two! A diagnosis, prosthesis, pacemaker or anti-hypertension treatment will always be goods, with their technical characteristics. However, the health specialist should always be in the service of the person being treated. This person has experience, feelings and a specific life project. Quality of life reinforces that the older patient is a person and thereby, in a certain way, helps make sure that the radical distinction found in the

Civil Code between people and goods is more fully acknowledged in the field of health.

QUALITY OF LIFE INDIRECTLY TAKES INTO ACCOUNT THE EMOTIONAL INTELLIGENCE OF THE PATIENT

However, all new ideas have their detractors. I found three types of attack against the credibility of assessing the quality of life with regard to health:

- quality of life is like love, beauty and happiness. It can't be measured;
- if you could define quality of life, you would be able to quantify it;
- if you are able to measure the quality of life, the way that you did so is faulty.

We can put these attacks into perspective by recalling an old debate about how to measure intelligence. Binet said: "Intelligence is what my test measures." We now know that there are types of intelligence that the Binet-Simon test doesn't take into account. For Victor Hugo; "Reason is intelligence in action; imagination is intelligence in erection!" In addition, the American psychologist Daniel Goleman (4), in his recent bestseller, *Emotional Intelligence*, made some comments that should be noted here. Goleman, considering the advances of knowledge in biology and neuropsychology, notes that our idea of intelligence is overly based on rational intelligence, the intelligence measured by the I.Q. test. Emotional intelligence involves motivation, self-control and social ease. In short, emotional life.

For Goleman, in practice, modern medicine has undertaken the task of treating the disease, but has neglected the emotional component. In addition, the medical model almost entirely rejects the idea that the mind can have an effect on the body. An assessment of the quality of life of the older patient may thereby help progressively change this *a priori* of medicine, since the quality of life indirectly takes into account the emotional intelligence of the patient.

Quality of life includes these different meanings of intelligence, which is multidimensional. Therefore, quality of life as related to health includes the combination of a generic (or general) instrument of evaluation that reflects physical, mental and social well-being and a specific instrument (the elements that reflect the assessed disease, for example angina, asthma, adenoma of the prostate or even urinary incontinence).

Quantity intrudes in the process of attempting to measure the quality of life. In fact, it does not primarily involve a measurement, but rather an assessment, since there are subjective aspects. It should be kept in mind that measurement does not allow for criticism as regards physical quantification, for example, the chronometer that times skiers during a downhill race or slalom to the nearest 1/100th of a second. In this case, the measurement is independent of the measurer and the data is not subject to interpretation. Assessment involves an estimate in which more approximate elements of judgement are found. This is the case of the scores to the nearest 1/10th of a point by each of the judges in a figure skating competition. In assessment, the data is subject to interpretation. In the field with which we are dealing, we can measure the difference for a subject, or a group of subjects, before and after a treatment, or any preventive or curative action. The measurement is then derived from two (or several) successive assessments on a quality of life scale. Quality of life isn't constant. It is often transient, especially within the realm of longevity.

HEALTH, LONGEVITY AND QUALITY OF LIFE

Health, longevity and quality of life represent a very contemporary medical and social combination. Health has been a supply market for a long time. What is changing today is that it is becoming (and will become) a demand market. The value of the service that the health professional gives the patient is of immaterial rather than of material value. The emergence of the patient is contemporary, although slightly out of step with the evolution of the consumer in post-industrial society.

The patient is accepting less and less being the user or being "serviced" in the system of health care. He wants more information about what concerns him. He wants the clinician to treat him as a subject and not an object suffering from a disease. He wants to be treated as an individual with a life project and to be active in the decisions that concern him. He wants to be treated as a person and not as goods, in application of the basic principles of the Civil Code, as we have already mentioned. This may finally lead to a discovery of certain values that have been up to now dissimulated by our technical and quantitative era.

Health, longevity and quality of life represent a very contemporary medical and social combination. The assessment of quality of life helped advance the clinics and the development of new treatments. The instruments for assessment are not fully developed. Their adaptation and application to a human life extended by progress, represents a new field of knowledge. However, the end of life can't be summed up by problems of memory, pain or sundry disabilities. It also involves the inevitable approach of the end of a human being and the consubstantial spiritual aspect. The paradox is that this spiritual dimension is arising in a wealthy materialistic society that has provided both increased longevity and decreasing demography, as well as the collapse of the Judeo-Christian beliefs and values upon which the Occident was built.

REFERENCES

1. Source: PNUD (Programme des Nations Unies pour le Dèveloppement) [1996], Rapport mondial sur de dèveloppement humain, Paris, Economica, 251 p.

2. Attali, Jacques, Dictionnaire due XX!e siėle, Paris, Fayard, p. 320.

3. Sachs, Lisbeth [1996], “Le risque comme diagnostic: retentissement sur la qualitè de vie”, Cancer, Sida, La Qualitè de Vie, under the direction of C. Jasmin, J.A. Levy and G. Bez, Paris, Les Empêcheurs de Penser en rond, 1996, 212 p.

4. Goleman, Daniel [1997], L’intelligence èmotionnel (Emotional Intelligence), Paris, Robert Laffont, pp. 212-213.

2. Ethics of Health Care for the Elderly

LONGEVITY, AGING AND THE DEMAND OF PRIMARY CARE

Chris van Weel

University of Nijmegan
Department of General Practice
and Social Medicine
P.O. Box 9101
6500 HB Nijmegen
The Netherlands

INTRODUCTION

General practice/family medicine provides medical care for individuals and their families in the community. A basic responsibility is the targeting of care to the specific needs and demands of the community that is served. The definition of the discipline of general practice emphasizes the importance of norms and values, the specific (1,2) epidemiology and medico-social interface.

Insight into changes in morbidity in the target population of care is essential, for which surveillance systems provide crucial information.

The pattern of morbidity in the population is determined by two factors. First, the frequency of a disease can change over time, for example, the increase of ischemic heart diseases in the Western world in the decades after the Second World War, and the appearance of HIV/AIDS in the late 1970s. In these instances, an increase in the incidence of disease can be explained by cigarette smoking and specific dietary habits in the former, and the pathogenicity of the HIV virus in the latter. A second factor is the variability of the population. Due to changes in gender sex, age or racial distribution the profile of diseases may rise or decline, even while the incidence of the disease does not change.

Aging and longevity characterize affluent Western society, and this process is reflected in the current and future demands for medical (3) and social services.

Longevity and Quality of Life, Edited by Butler and Jasmin.
Kluwer Academic / Plenum Publishers, New York, 2000.

There is a strong tendency to project clear negative consequences for the health care budget, despite evidence to the contrary (4,5).

This paper analyses the effects of the aging population on the epidemiology and workload of general practice. It scrutinizes the generally held view of aging as an indiscriminate factor in the increasing demand for health care.

METHODS

The study is based on general practice morbidity data from the Continuous Morbidity Registration (CMR) of the Department of General Practice and Social Medicine, University of Nijmegen. The study was based on the most common morbidity in general practice. Two analyses were made:

- Trends over time of incidence and prevalence between 1971 - 1997.
- Extrapolation of disease frequency in the predicted aging of the Dutch population (6).

THE DATABASE: GENERAL FEATURES CMR (7)

The Continuous Morbidity Registration Nijmegen is a network of four general practices in the Nijmegen region (eight general practitioners). The network surveys and monitors morbidity in general practice on an ongoing basis since 1971 (in two practices since 1967). All new episodes of illness are recorded, including diagnoses made after referral. The database is used for analysis of general practice morbidity and morbidity trends since 1971, and as an index for the recruitment of groups of patients for additional research.

The practice population is about 12,000 patients and has been stable over the years. The relevance and the limitations of the register are directly influenced by the Dutch health care structure. Two aspects of this structure are of particular importance in this respect: in the Dutch health care system, the general practitioner has a 'fixed' list of patients (the practice population), and he/she is the gate keeper of the access to professional medical care. As a consequence, the system collects all morbidity in a defined population for which professional medical care is sought.

Each episode of morbidity presented to the general practitioner is recorded including the cause of death. The general practitioner diagnosing the episode provides the diagnostic codeing. An episode of morbidity is defined according to the international glossary for family practice (8): Follow-ups of already recorded morbidity are not recorded.

The morbidity data are stored by date of presentation/diagnosis, in relation to the demographic data of the patient: sex, age, social class and family composition. The registration started in 1967, and has been uninterrupted since that time. The population of the four practices has been remarkable stable over the years, and as a consequence it is possible to analyze long-term individual morbidity. As longitudinal studies of morbidity in general practice have become the main objective of the register, the efforts to secure the validity of the data must be seen in this light.

INTERNAL QUALITY CONTROL OF DATA

Quality control of recorded data involves the following activities:

Diagnostic Classification and Diagnostic Definitions

At the start of the register in 1967, the only morbidity classification for general practice available was the Dutch translation (9) of the British E-list (10). For the sake of consistency, over time the classification has never been changed when classifications more suitable for general practice became available (11) . But the list has been made compatible with the International Classification of Health Problems in Primary Care (ICHPPC) (11), and the ICHPPC definitions have been introduced.

Procedure of Classifying and Coding

Each episode of morbidity is classified and coded by the general practitioner who is involved in the case. This should be done as soon as possible after the consultation, and must reflect the highest level of diagnostic interpretation of the patient's condition. But in case of uncertainty, the classification/coding may be postponed until more certainty has been achieved (from the natural history of the disease, diagnostic tests or specialists' assessment).

Training and Support of General Practitioners

All general practitioners involved in the register have been trained in the use of the classification list and the application of the ICHPPC definitions (using vignette cases). There is a monthly meeting of all general practitioner to discuss coding problems and to monitor the application of diagnostic criteria. The practice assistants are trained and supervised in collecting the demographic data of the patients.

Completeness of Data

The practice assistants supervise the transfer of the coded data to the Department of General Practice and Social Medicine, where analysis takes place. By comparing the patient's files and coded diagnoses, the assistants monitor the completeness of the data.

DEFINITIONS

The definition of old age is in itself an ambiguous one. Traditionally, the age of retirement – 65 – is used to identify *the elderly*. A more realistic, but equally disputable cut-off point is the age of 75. When discrete age classes are presented in this study, both are used.

Incidence is the annual number of newly identified cases of a disease per 1,000 individuals in the practice population, standardized for sex and age.

Prevalence is the annual number of all identified cases of a disease per 1,000 individuals in the practice population, standardized for sex and age.

The most common acute (12) and chronic (7,12) diseases encountered in (primary) medical care were considered (Tables 1a and 1b).

USE OF RESOURCES AND REFERRALS

The CMR database does not provide information on the use of general practice facilities (number of consultations and home visits). These data have been calculated from the CMR incidence/prevalence and the consultation/visit rates found in the Dutch National Study on morbidity and interventions in general practice (13,14). The use of hospital specialist facilities has been calculated from the CMR recorded referrals.

RESULTS

The recorded morbidity concerns all body systems and organs, as can be expected from a general practice that specializes in professional care for unselected health problems. As can be identified from Tables 1a and 1b, common acute illness is frequently encountered at all ages. Chronic diseases are more common with increasing age, but are in no way specific for the elderly. An important feature with increasing age, however, is co-morbidity or multi-morbidity: the simultaneous presence of two or more chronic conditions in the same individual (7).

In the trends over time, the *incidence* of most chronic diseases has remained stable, whereas the *prevalence* has more than doubled between 1971 and 1995 (7). When extrapolating the morbidity frequency for the period 1996 - 2050, most chronic diseases (neoplasms, diabetes mellitus, chronic arthritis, COPD, cardiovascular diseases) show an increase as illustrated for cardiovascular diseases. The general pattern is that of a clear increase for the period 1996-2030, with leveling-off after that time. However, neonatal conditions, pregnancy-related care and asthma, gradually lose prominence.

For the use of primary and secondary care facilities the increased trend follows the same pattern, with the strongest increase in the period to 2030, and a leveling-off until 2050. Use of general practice facilities increases by 16 percent by the year 2050—mainly due to an increased demand for home visits. Though this concerns the general practitioner, there is, in particular, an increased demand for tasks that can be handled by the practice assistant. Referrals to geriatricians, radiotherapists and cardiologists are expected to increase (Table 2), but referrals to pediatricians, gynecologists and ENT surgeons is expected to decline.

CONCLUSION

From the presented data it can be concluded that the frequency of diseases, and the related demand for primary and secondary care, will change in the coming decades. In analyzing the contributing factors, these changes follow mainly from population demography: the incidence of the most common chronic conditions encountered in general practice do not appear to increase (7). Insofar as the extrapolations are a correct prediction, most of the change will take place in the coming three decades, and will level-off afterwards. It is important to acknowledge this relative short-term effect in the planning of health care facilities. It is also important to realize it is a change rather than an outright increase in needs and demands: the pediatrician and obstetric/gynecologist symbolize an era of youthful populations of the past, and the decline in the demand of their services will be replaced by that of radiologists and geriatricians, heralding an aging of the population. Of particular importance is the emphasis on comorbidity/multimorbidity in an aging population (7). This will require support from community-based medical services: an increase of the workload of general practitioners, but even more of medical assistants.

An interesting finding is the marked increase in the prevalence of chronic diseases in general practice, with their incidence staying constant. The cause of this increase is not directly obvious – a possible explanation is the increased life expectancy due to more effective treatment, in that the same number of patients will live longer after diagnosis. But this seems improbable when one considers the short period within which this phenomenon has taken place. More likely, it reflects a change in general practice routines. Greater emphasis on pro-active and continuous care of patients with chronic conditions (15) has resulted in patients receiving closer follow-up. This may indicate the strong influence of (changes in) medical routines on the demands for health care facilities, a factor that can profoundly influence the predictions presented in this paper.

This study has focused on illness and disease in relation to the aging of the population. Although there is a clear correlation, it is important to keep in mind that longevity and aging are powerful markers of the health of a population. Elsewhere we have hypothesized that terminal illness is an important factor in the costs of health care (12) – resources spent on palliation and support of individuals during their final stage of life. As more and more individuals live to great old age, there is a danger that these costs will be attributed to old age, and is used to stigmatize 'the elderly'.

Rather than worry about some of the inherent consequences for health care, society should realize the potentials of counting among its members increasingly large numbers of strong and healthy older people.

Table 1A

Age specific incidence of most common primary care morbidity.
Incidence: number of new cases/1,000 patients practice list.
Source: Nijmegen Continuous Morbidity Registration 7, 1991-1995

Illness	75+ y.	65-74 y.	45-65 y.
Common Cold	210	194	153
Urinary Tract Infection	149	99	40
Ear Wax	126	106	59
Bruse, Contusion	102	41	32
Dermatitis	86	78	61
Psychosomatic Complaints	53	52	86
Obstipation	53	19	10
Myalgia	48	53	66
Acute Bronchitis	47	30	18
Low Back Pain	35	39	47

Table 1B

Age specific prevalence of most common primary care chronic diseases.
Incidence: number of new cases/1,000 patients practice list.
Source: Nijmegen Continuous Morbidity Registration 7, 1991-1995

Disease	75+ y.	65-74 y.	45-64 y.
Osteoarthritis hip/knee	326	163	35
Deafness	254	127	30
Obesity	218	204	150
Hypertension	207	233	114
Cataract	196	60	6
Heart Failure	149	44	5
Chronic Ischemic Heart Dis.	148	110	27
COPD	111	105	36
Diabetes Mellitus	109	84	27
Stroke	100	44	9

Table 2

Expected number of referrals in a standard general practice.
Percent increase from 1996
Source: CMR based data

	1996	2020	% +	2035	% +	2050	% +
General physician	18	21	19	23	32	24	37
Pulmonology	7	8.2	21	9	28	9	28
Cardiology	17	21	27	23	39	24	42
Rheumatol.	2	3	9	3	17	3	17
Pediatricics	8	7	-18	7	-17	7	-18
Surgery (General)	39	43	10	46	17	47	19
Surgery (Cardiovasc)	2	2	27	2	40	2	33
Surgery (Plastic)	7	7	1	7	-1	7	0
Surgery (Oropharynx)	2	2	0	2	-6	2	-6
Urologist	14	15	7	16	14	16	15
Surgery (Orthopedic)	26	28	6	28	8	28	9
Surgery (Pediatric)	2	1	-18	2	-12	1	-18
Radiology	0	1	50	1	75	1	75
Surgery- (Eye)	35	42	19	46	31	46	33
Surgery (ENT)	27	28	4	30	8	30	8
Dermatology	20	20	3	20	3	20	4
Gynecology	25	22	-12	22	-14	21	-15
Rehabilitation	3	3	12	3	24	3	28
Geriatrics	1	1	50	1	83	1	100
Psychiatry	4	4	3	4	-3	4	-3
Neurology	17	20	13	21	20	21	21
Total	566	589	4	604	7	608	7

REFERENCES

1. KNAW (Koninklijke Nederlandse Academie van Wetenschappen) General Practice Research in Dutch Academia, Amsterdam, 1994.

2. Donaldson, M., Yordy, K. and Vanselow, N. editors. Defining primary care: an interim report. Institute of Medicine, Washington DC, 1994.

3. Ruwaard, D. and Kramers PGN (eds). Volksgezondheid Toekomst Verkenning 1997 - de som der delen. Elsevier/de Tijdstroom, Utrecht, 1997.

4. Perls TT, Wood ER. Acute care costs of the oldest old: they cost less, their care intensity is less and they go to nonteaching hospitals. Arch Intern Med 1996; 156: 754-60.

5. Greenberg, I. and Cultice, JM. Forecasting the need for physicians in the United States: the Health Resources and Services Administration's physician requirements model. Health Serv Res 1997; 31: 723-37.

6. Vademecum Health Statistics of the Netherlands 1996. Statistics Netherlands Ministry of Health, Welfare and Sports, Rijswijk, 1996.

7. van Weel C., Chronic diseases in general practice: the longitudinal dimension. Eur J Gen Pract 1996; 2: 17-21.

8. WONCA. An international glossary for primary care. J Fam Pract 1981; 13: 671-81.

9. Anonymous. Morbidity classification for general practitioners derived from English E-list. [Morbiditeitsclassificatie voor gebruik door huisartsen, ontleend aan de Engelse z.g. 'E-lijst']. Utrecht: Dutch College of general practitioners, 1963.

10. Eimerl, T.S., Organized curiosity. A practical approach to the problem of keeping records for research purposes in general practice. J Col Gen Pract 1960; 3: 246-52.

11. Classification Committee of WONCA in collaboration with WHO. ICHPPC-2 defined (International classification of health problems in primary care). 3rd edition. Oxford: Oxford University Press, 1983.

12. van Weel, C. and Michels, J., Dying, not old age, to blame for costs of health care. Lancet 1997; 350: 1159-60.

13. Groenewegen, P.P., de Bakker DH, van der Velden J. Een Nationale Studie naar ziekten en verrichtingen in de huisartspraktijk. Basisrapport, Verrichtingen in de huisartspraktijk. Nederlands Instituut voor onderzoek van de Eerstelijnsgezondheidszorg (NIVEL), Utrecht, 1992.

14. FÜrst MEG. De huisartspraktijk na het jaar 2000. Verslag wetenschappelijke stage. Nijmegen, KUN, FMW, 1998.

15. Hart, J.T., Reactive and proactive care: a crisis. Br J Gen Pract 1990;40: 4-9

HEALTH INEQUALITIES PERSIST INTO OLD AGE: RESULTS FROM THE LONGITUDINAL STUDY

Elizabeth Breeze

London School of Hygiene and Tropical Diseases
Keppel Street
London WC1E 7HT
United Kingdom

INTRODUCTION

Health inequality is here defined as a systematic difference in health outcomes between socioeconomic groups, especially an association between poorer social and economic circumstances and greater morbidity and mortality. The debate on socioeconomic inequalities in health outcomes has focused mainly on premature mortality, before the age of 65 years. Less attention has been paid to older people, in part because of difficulties in assigning socioeconomic status to older people. Other important health outcomes in older people include institutional care and chronic ill health. We used the Longitudinal Study (LS) to investigate the risk for people in middle age and early retirement for subsequent mortality, risk of institutional care and limiting long-term illness. We were also interested in whether changes in circumstances as people moved into retirement and through old age predicted health outcomes. In this report these three outcomes are discussed in terms of housing tenure and availability of a car to the household as measures of material conditions and socioeconomic status.

THE LONGITUDINAL STUDY

This comprises one percent of people recorded on the 1971 Census in England and Wales, enhanced by subsequent births and immigration. The data for these individuals from successive Censuses (1981 and 1991) is recorded together with vital events, including death. The data set is maintained by the Office for National Statistics and the Social Science Research Unit of City University, London. (1)

Longevity and Quality of Life, Edited by Butler and Jasmin.
Kluwer Academic / Plenum Publishers, New York, 2000.

METHODS

Analyses were undertaken separately for men and for women and for two ten-year cohorts, aged 55-64 years and 65-74 years, respectively, in 1971. Mortality analyses included deaths up to the end of 1992, the last date for which data were available at the time of analysis. Institutional residence (predominantly residential homes and nursing homes in which individuals receive substantial care) was recorded at each Census and limiting long-term illness was first recorded at the 1991 Census. A person was designated as having a limiting long-term illness if the answer was "yes" to the question: "Do you have any long-term illness, health problems or handicap which limits your daily activities or the work you can do? (Include problems which are due to old age)."

Besides housing tenure and car availability, the Census variables used as predictors were occupational social class, marital status and whether living alone. Housing tenure and car availability are attributes of the household in which the individual lives. To be in an owner-occupied accommodation the subject need not be the owner and to be in a household with a car available the subject need neither own nor use the car. The results that are presented are adjusted for all the above factors and also for five-year age groups within cohorts. Mortality models further include five-year time periods for year of death. Time adjustments were not possible for institutional care and limiting long-term illness as these were only recorded at the time of the Census.

Mortality over a 21-year period was analyzed by 1971 circumstances for all those aged 55-64 in 1971, including those already in institutions. Mortality over the period 1981-1992 was analyzed by changes in circumstances between 1971 and 1981 for those who were in the community in 1971 and 1981 (otherwise socioeconomic changes could not be measured). Institutional residence in 1991 was analyzed by 1971 circumstances and by inter-Census changes, in both cases for those who were living in the community in 1971 and 1981 and alive in 1991. Limiting long-term illness was analyzed for those who were additionally still in the community in 1991.

RESULTS

Description of the Population (Table 1)

There were 43,092, men and 50,839 women aged 55-74 years in the LS sample in 1971. Over 80 percent of men were married, around 70 percent of women aged 55-64 and nearly 50 percent of women aged 65-74. Small proportions of men were widowed (4 percent of the younger group, 12 percent of the older group) but more substantial proportions of women (17 percent of younger women, 36 percent of older women). Overall, 29 percent lived in owner-occupied accommodation with household access to a car (considered to be the most favorable circumstances) and 34 percent lived in rented accommodation without household access to a car (the least favorable circumstances). We attributed the social class of her husband to a woman who was married in 1971. The populations analyzed for institutional residence and limiting long-term illness excluded those who had died before 1991 or had moved into an institution in 1981. They were more predominantly people in owner-occupied accommodation with household access to a car and less likely to be in the less privileged circumstances than the full 1971 population

By the end of 1992, 70 percent of the younger cohort of men and 49 percent of the women had died, as had 93 percent of the older men and 84 percent of older women. Institutional residence was more likely to be experienced by women than by men, with 6 percent of younger women and 3 percent of younger men in an institution in 1991; the equivalent percentages for the older cohort were 23 percent and 14 percent. Of those who were in the community in all three Census years nearly half the younger cohort reported a limiting long-term illness in 1991 (44 percent men, 47 percent women) and nearly two-thirds of the survivors in the older cohort (65 percent men, 60 percent women). By 1991 the survivors of the younger cohort were aged 75-84 and those of the older cohort aged 85-94.

Relative Risks of Health Outcomes by Changes in Housing Tenure and Car Availability (Table 3)

For the younger cohort the relative risk of mortality, institutional residence and limiting long-term illness were all significantly raised if they were in rented accommodation or in a household without access to a car in 1971. The additional risk for those with both disadvantages was in the order of 40 percent for mortality, of 90 percent for being in an institution if male and 50 percent if female, and 20-30 percent for suffering a limiting long-term illness. There was also some additional risk if only one of these circumstances applied, with the exception of women in rented accommodation, with a car available. For the older cohort the results were much weaker but still statistically significant for mortality. While only car availability had a noticeable association with institutional residence and housing, neither tenure nor car availability in 1971 were associated with presence of a limiting long-term illness among the very old in 1991.

Description of Changes in Circumstances Between 1971 and 1981

By 1981, 40 percent of men and 26 percent of women in the 1971 sample had died. The mortality analyses for inter-Census changes were based on 25,758 men and 37,174 women who had completed the Census form in 1971 and 1981 and were living in the community at both Census points. The analyses of institutional residence and limiting long-term illness by inter-Census change were based on the same sub- populations of survivors as the analyses by 1971 circumstances.

The most common changes experienced were loss of spouse and starting to live alone. Nearly 1 in 4 women and 1 in 10 men experienced loss of spouse and slightly smaller percentages started to live alone. Between one in six and one in four subjects lost household availability of a car (depending upon the age and gender) and about half as many gained it. Four to six percent moved out of owner occupation into rented accommodation and smaller numbers moved in the reverse direction. The profile of changes was fairly similar for all those in the community in 1971 and 1981 and for the subset who survived until at least 1991.

Relative Risks of Health Outcomes by Changes in Housing Tenure and Car Availability Between 1971 and 1981 (Table 3)

Again, for the younger cohort there was a clear increase in risk of mortality, institutional residence and limiting long-term illness if they were either in rented accommodation in both Census years or in a household without a car available in both years. These are assumed to be households with long-term material disadvantage. A change from owner-occupation to rented accommodation was associated with increased risk of mortality and, for women, of living in an institution. Losing availability of a car was associated with a modest increased risk of all three outcomes for men and women but the increase was greater for men than for women. Changes which were considered to be improvements (i.e., into owner-occupation or gaining access to a car) were not necessarily associated with lower risks than for those who did not experience the improvements. For the older cohort the small numbers reduced the power to detect differences in risk but associations between change and health outcomes were also weaker. However, women who moved out of owner occupation were at greater risk of all three outcomes (26 percent more likely to die over an 11-year period than those who did not) and men who ceased to have household access to a car were 36 percent more likely to die than those who did not. Also, lack of a car in both Censuses was associated with some increased risk of all three outcomes in the older cohort.

DISCUSSION

Long-term follow-up studies can provide unique insights into risks associated with living patterns at an earlier age. The additional advantage of the LS is the linking of successive Censuses which provide a better picture of individuals' circumstances as they move through life than a measurement taken at a single point in time.

These results show the persistence of unfavorable socioeconomic indicators in middle age and early retirement on long-term outcomes in old age. Moreover, loss of favorable socioeconomic indicators—owner occupation and household availability of a car—during retirement was also associated with increased risk of 11-year mortality and institutional residence or limiting long-term illness 10 years later. There is insufficient evidence to say without doubt that the material circumstances were directly implicated in worsening health; for example, there were no measurements of health behaviors or supply and use of health services. However, while lifestyle factors are still relevant at least in early old age (2, 3) they would not necessarily account for all the association between socioeconomic factors and mortality (4) or morbidity (5); moreover they may themselves be consequences of socioeconomic and demographic circumstances.

It can also be argued that poor health led to poor socioeconomic circumstances rather than vice versa. We re-ran the mortality analyses, excluding deaths that occurred in the first three years, on the assumption that if poor health had led to the loss of owner occupation or household availability of a car then there would be a high rate of mortality in the early years among those who had experienced this change, and in later years there would no longer be clear differences between the socioeconomic groups.

After excluding three years, the relative risks of 21-year mortality for those in institutions in 1971 were reduced but still significant. It may be that those who enter institutions relatively young have chronic diseases which are less immediately life- threatening. There were still statistically significantly increased risks of 11-year mortality for moving out of owner-occupation for younger men and older women and excess risks associated with losing household availability of a car for men and for younger women. Thus, health selection did not explain all the excess risk. However, it did explain excess risk of 11-year mortality associated with changing to a household with a car available. For example, sick subjects may have moved in with a son or daughter (with a car) who could care for them. Similar checks could not be carried out for institutional residence and limiting long-term illness but it seems unlikely that these people would survive in 1991 if poor health had led to deterioration in circumstances by 1981.

The LS does not include data on institutional admissions between Census points and therefore individuals in institutions who died before the Census would be excluded. The average length of stay in a home is 2-3 years and if people stay long enough to become residents they are likely to die there. The relative risks of entering an institution for the socioeconomically disadvantaged will be overestimated if those who entered institutions and died before the Census were disproportionately from the advantaged groups, e.g., if the advantaged groups were either more disabled than the disadvantaged in 1971 or deteriorated more rapidly. Both scenarios seem unlikely. On the other hand, the relative risks will be underestimated if those whose institutionalization was missed were more likely to be from the disadvantaged groups. As mortality was higher in these groups, this seems more feasible.

Generally, there were fewer statistically significant results for the older cohorts than the younger ones, partly because of the reduced statistical power from smaller numbers. Many of the relative risk estimates were close to 1.00 suggesting attenuation of differences as people survive into very old age. Given the advanced age of the older cohort in 1991 it is remarkable that there were any associations between health outcomes and socioeconomic factors.

The choice of socioeconomic indicator for older people is problematic. Occupational social class is not reported here because the pattern of risks was not clear. There is considerable scope for misclassification; the information may have been provided by someone else in the subject's household and refers to a job from which the subject may have been retired many years. Housing tenure and car availability are less prone to misreporting but the results of the analyses by change in circumstance suggest that it is insufficient to measure these factors only at one point in time.

In summary, these results demonstrate the persistence of inequalities in health-related outcomes throughout old age, both in those with unfavorable circumstances in mid-life and in those who, in later life, have lost earlier advantages.

Table 1

Baseline characteristics of A) all men and women in the 1971 LS sample, B) those in the community in 1971 and 1981 and alive in 1991.

Characteristic in April 1971	MEN				WOMEN			
	55-64 in 1971		65-74 in 1971		55-64 in 1971		65-74 in 1971	
	A	B	A	B	A	B	A	B
HOUSING TENURE AND CAR AVAILABILITY	%	%	%	%	%	%	%	%
In owner-occupation , car	36	45	27	38	31	35	20	25
In owner-occupation, no car	15	13	25	24	20	20	30	31
In rented accommodation, car	20	21	10	12	15	16	7	7
In rented accommodation, no car	27	21	36	26	32	29	41	37
In institution	1	-	2	-	1	-	2	-
SOCIAL CLASS*	%	%	%	%	%	%	%	%
I/II	22	27	18	24	18	20	13	17
IIIM	11	11	11	15	13	14	9	11
IIIN	33	34	27	27	24	25	14	16
IV/V	29	25	31	28	26	25	19	20
Unclassified†	6	2	12	6	19	16	44	36
N	26222	8947	16870	1517	28383	15458	22895	4604

* Social Class derived from occupation at the Census or most recent occupation.
Currently married women have been assigned the social class of their husband, other women their own social class.
†Those who could not be assigned a class either because of inadequate information or because they did not have an occupation

Table 2

Risk ratio (RR), 95% confidence intervals (CI) for health outcomes by socioeconomic circumstances in 1971.

Characteristic in 1971	55-64 years in 1971			65-74 years in 1971		
	Deaths 1971-92	Institution 1991	Long-term illness 1991	Deaths 1971-92	Institution 1991	Long-term illness 1991
MEN	Adjusted† RR 95% CI	Adjusted† RR 95% CI	Adjusted† RR 95% CI	Adjusted† RR 95% CI	Adjusted† RR 95% CI	AdjustedH RR 95% CI
TENURE AND CAR AVAILABILITY						
Owner occupation, with car	1.00	1.00	1.00	1.00	1.00	1.00
Owner occupation, no car	1.26 *1.19-1.32*	1.54 *1.05-2.26*	1.20 *1.11-1.29*	1.19 *1.13-1.24*	1.73 *1.15-2.59*	1.04 *0.91-1.16*
Renting, with car	1.21 *1.16-1.27*	1.47 *1.03-2.09*	1.10 *1.02-1.17*	1.12 *1.05-1.19*	1.24 *0.72-2.12*	0.96 *0.80-1.11*
Renting, no car	1.44 *1.38-1.50*	1.89 *1.34-2.67*	1.22 *1.14-1.31*	1.36 *1.31-1.42*	1.94 *1.28-2.94*	1.09 *0.96-1.21*
Communal	1.25 *1.07-1.47*	Not applicable	Not applicable	1.59 *1.37-1.81*	Not applicable	Not applicable
WOMEN						
TENURE AND CAR AVAILABILITY						
Owner occupation, with car	1.00	1.00	1.00	1.00	1.00	1.00
Owner occupation, no car	1.15 *1.09-1.22*	1.26 *1.03-1.55*	1.12 *1.05-1.18*	1.14 *1.08-1.19*	1.25 *1.02-1.54*	1.03 *0.96-1.10*
Renting, with car	1.15 *1.09-1.22*	1.01 *0.79-1.28*	1.19 *1.12-1.26*	1.14 *1.06-1.22*	0.98 *0.71-1.36*	1.02 *0.92-1.12*
Renting, no car	1.38 *1.32-1.45*	1.45 *1.20-1.75*	1.26 *1.20-1.32*	1.33 *1.27-1.39*	1.39 *1.13-1.71*	1.09 *1.03-1.15*
Communal	1.81 153-2.12	Not applicable	Not applicable	2.67 *2.39-2.96*	Not applicable	Not applicable

†Adjusted for five-year age bands, marital status, whether living alone and social class; mortality also adjusted for five-year time bands

Table 3

Risk ratio (RR), 95% confidence intervals (CI) for health outcomes by socioeconomic circumstances between 1971 and 1981.

Change in characteristic	55-64 years in 1971			65-74 years in 1971		
MEN	Deaths 1981-92 Adjusted RR † 95% CI	Institution 1991 Adjusted RR † 95% CI	Long-term illness 1991 Adjusted RR † 95% CI	Deaths 1981-92 Adjusted RR † 95% CI	Institution 1991 Adjusted RR † 95% CI	Longterm Illness 1991 Adjusted RR † 95% CI
HOUSING TENURE						
In owner occupation 1971, 1981	1.00	1.00	1.00	1.00	1.00	1.00
Moved:						
to owner-occupation	1.07 *0.98-1.17*	1.09 *0.64-1.87*	0.98 *0.87-1.09*	1.02 *0.88-1.17*	2.04 *1.02-4.06*	0.95 *0.69-1.19*
out of owner-occupation	1.19 *1.06-1.33*	1.20 *0.61-2.39*	1.12 *0.95-1.29*	1.04 *0.91-1.19*	1.29 *0.64-2.59*	1.20 *0.95-1.40*
In rented accommodation	1.07 *1.15-1.26*	1.42 *1.06-1.89*	1.09 *1.02-1.15*	1.04 *0.97-1.10*	1.06 *0.75-1.50*	1.04 *0.92-1.15*
CAR AVAILABILITY						
Available in 1971& 1981	1.00	1.00	1.00	1.00	1.00	1.00
Gained availability	1.20 *1.09-1.31*	1.34 *0.75-2.39*	1.09 *0.96-1.21*	1.26 *1.12-1.43*	1.07 *0.47-2.44*	1.05 *0.80-1.25*
Lost availability	1.43 *1.34-1.52*	1.56 *1.03-2.37*	1.26 *1.16-1.36*	1.36 *1.25-1.47*	1.18 *0.66-2.09*	1.08 *0.91-1.24*
No car in either year	1.32 *1.26-1.39*	1.61 *1.18-2.18*	1.24 *1.17-1.31*	1.27 *1.20-1.35*	1.77 *1.21-2.60*	1.08 *0.97-1.19*
WOMEN						
HOUSING TENURE						
In owner occupation 1971, 1981	1.00	1.00	1.00	1.00	1.00	1.00
Moved:						
to owner-occupation	1.06 *0.97-1.18*	0.86 *0.60-1.23*	1.16 *1.06-1.25*	1.00 *0.89-1.12*	1.03 *0.73-1.46*	0.89 *0.75-1.03*
out of owner-occupation	1.17 *1.05-1.31*	1.73 *1.28-2.33*	1.11 *0.99-1.22*	1.26 *1.14-1.39*	1.43 *1.02-2.00*	1.20 *1.05-1.34*
In rented accommodation	1.22 *1.16-1.28*	1.23 *1.06-1.43*	1.15 *1.10-1.20*	1.14 *1.08-1.19*	1.10 *0.94-1.29*	1.08 *1.02-1.15*
CAR AVAILABILITY						
Available in 1971& 1981	1.00	1.00	1.00	1.00	1.00	1.00
Gained availability	1.06 *0.97-1.19*	1.20 *0.84-1.72*	0.98 *0.89-1.08*	1.04 *0.93-1.17*	0.99 *0.68-1.44*	0.95 *0.80-1.09*
Lost availability	1.13 *1.05-1.21*	1.36 *1.07-1.73*	1.12 *1.05-1.19*	1.01 *0.92-1.10*	1.04 *0.78-1.38*	1.06 *0.95-1.16*

ACKNOWLEDGMENTS

The project was funded by the Department of Health, UK. The Longitudinal Study Team at the Office for National Statistics and The Research Unit Team at City University made available the LS Data. The views expressed in this paper are not necessarily those of the Office for National Statistics or the Department of Health.

REFERENCES

1. OPCS Longitudinal Study. User Manual. London: City University, 1990

2. Kaplan GA, Seeman TE, Cohen, RD, Knudsen, LP, Guralnik, J., Mortality among the elderly in the Alameda County Study: behavioral and demographic risk factors. Am J Publ Health 1987;77:307-312.

3. Davis M, Neuhaus JM, Moritz DJ, Lein D, Barclay, JD, Murphy, SP., Health behaviors and survival among middle-aged and older men and women in NHANES I Epidemiologic Follow-Up Study. Prev Med 1994;23:369-376.

4. Davey Smith, G, Shipley, MJ, Rose, G., Magnitude and causes of socioeconomic differentials in mortality: further evidence from the Whitehall Study. J Epidemiol Community Health 1990;44:265-270.

5. Cairney, J., Arnold R., Social class, health and aging: socioeconomic determinants of self-reported morbidity among non-institutionalized elderly in Canada. Can J Publ Health 1996;87:199-203.

GENTLE CLOSURE

Kathleen M. Foley

Project on Death in America
400 West 59th Street
New York, NY 10019
USA

INTRODUCTION

As we extol the benefits of the worldwide revolution in longevity and quality of life, we also need to acknowledge the obvious—the inevitability of death. Denial of death has significant consequences in preventing and limiting care of the most vulnerable of our citizens, the dying. As we focus on enhancing the quality of life for our aging population, special attention should be given to the quality of life for dying patients and their families. There is increasing awareness of the need to improve the care of the dying at both national and international levels. The experience of dying has changed over the past several decades with many more people enduring prolonged death as a consequence of chronic and progressive disease.

Needless suffering—physical, emotional, existential, and spiritual—too often accompanies these deaths for both dying patients and survivors. We are hindered in our efforts to improve the experience of dying for patients and families by a number of factors: deficiencies in many of our current models of caring for dying patients and their families; serious weaknesses in our education of health care professionals about care near the end of life, and major inadequacies in our knowledge about the course, treatment and outcomes of care of dying patients and their families.(1,2,3)

The Project on Death in America (PDIA) was created to begin to address these issues. Its mission is to understand and transform the culture and experience of dying through initiatives in research and scholarship, the humanities and arts; to foster innovations in provision of care, public education, professional education, and public policy.(4) The Open Society Institute, a nonprofit foundation created by George Soros that supports the development of open societies worldwide, funds the

Longevity and Quality of Life, Edited by Butler and Jasmin.
Kluwer Academic / Plenum Publishers, New York, 2000.

PDIA. Mr. Soros established the PDIA following his own experiences with the death of his parents.

From July 1994 to June 1998, the Project has provided funding for programs and individuals to address, through research and model systems of care, the barriers modern society and medicine face in providing appropriate compassionate care of the dying. Confounding these barriers are major medical, social and economic factors. These include: changes in the trajectory of dying, with large numbers of patients with, for example, cancer and AIDS, living for months or years following diagnosis; the increasing age of the population; advances in high technological medical support systems for patients with respiratory and cardiac failure; a greater emphasis on patient autonomy with the concerted shift from societal to individual rights; and the widely debated limitations in health care resources, particularly for patients with chronic medical illness.(2)

To date, the PDIA has played a unique role in developing initiatives and supporting programs that focus on providing appropriate competent, compassionate care for patients near the end of life. These initiatives emphasize the need to create a national agenda that places emphasis on several important goals:

- To better understand the dying process for patients and families,
- To develop health care professional expertise in palliative care and models of palliative care service,
- To create a national "death talk" to facilitate public discourse on what has been a private topic,
- To create a better reimbursement system for care of dying patients and their families,
- Achieving these goals will lead to a transformation in the culture of death in America.

The PDIA has created a Faculty Scholars Program, Grants Program and a series of Special Initiatives to improve care for the dying and their families and to raise public awareness and discussion about death and dying.

To address the needs to improve healthcare professional education, PDIA developed a Faculty Scholars Program to provide a cohort of clinicians, researchers and educators who are committed to work in this area. One of the goals of the Scholars Program is to promote the visibility and prestige of clinicians committed to this area and to enhance their effectiveness as academic leaders, role models and mentors for future generations of health professionals. To date, PDIA has supported 50 Faculty Scholars in 33 medical schools in the United States and 4 medical schools in Canada; one university-based nursing program and one community college-based nursing program. The faculty scholars include 46 physicians, 3 nurses and 1 Ph.D. These scholars are committed to institutionalizing change in their clinical practice, education and research environments. The faculty scholars work in various disciplines in medical ethics, medical education, economics, geriatrics, psychiatry, critical care, neurology, pediatric oncology, and general medicine and general nursing. Recognizing that the majority of Americans continue to die in institutions, either hospitals or nursing homes, creating role models for change in these institutions is critical to making an impact on the medical aspects of the dying process. This program seeks to provide participants with knowledge and skills necessary to develop innovative programs in clinical care, research, education and advocacy and to take leadership roles in their institutions and nationally.(5) The Faculty Scholars Program has developed an intellectually vibrant, mutually supportive and cross-fertilizing network of colleagues.

The PDIA Grants Program developed seven priority areas for funding and initially created four cycles of funding starting in January of 1995 through June of 1996. This process served to demonstrate the level of interest, expertise and need in these priority areas. During this period of time, PDIA received over 1,500 letters of intent and eventually funded 94 projects. These projects range in funding from $5,000 to $300,000 and all address important issues impacting on the care of the dying. They support bridging programs for families, in which patients dying from AIDS help to create foster care alternatives for their children; underwriting the Missoula, Montana Community Project to study death and dying in a community setting; supporting a novel home care program for homeless substance abusers dying of AIDS or cancer in Washington, DC; and a grant to the United Hospital Fund in New York City to develop a program to establish palliative care programs in 12 hospitals.

The seven priority areas include the epidemiology, ethnography and history of dying and bereavement in the United States. In this priority area, PDIA supported research and analyses of the social forces, past and present, which affect the process of dying, especially as they provide a foundation for change: where and how death occurs, and under what social and medical conditions; the services that are delivered to dying persons and their families, at what expense and under what types of legal and ethical guidelines; how different communities and different cultural, religious and socioeconomic groups respond to the process of dying and grieving; how health care professionals predict outcomes, including death and suffering, how the plan of care develops and how it unfolds.

In the second priority area—the physical, emotional, spiritual and existential components in dying and bereavement—the objective in this funding category was to improve the prospects for better outcomes: a gentle closure, reconciliation, physical comfort, and a coming to terms with a life that was led. The project supported both traditional and non-traditional approaches from within and outside the medical model: to help sustain the emotional and spiritual needs of dying persons and their families; to identify and respond to the unique needs of specific communities, such as those experiencing high rates of mortality from disease or violence; and to develop and implement strategies for incorporating spiritual and humanistic values into the dying and grieving process.

The contribution of the arts and humanities is a third priority area for PDIA and encourages persons from the literary, visual and performing arts to use their skills and insights to identify, create and convey meaning in facing disability and death; and to evoke and deepen our understanding of the diverse myths and metaphors that shape the experience of suffering, dying and bereavement.

A fourth priority was the development of new service delivery models for the dying and their families and friends. In this area, the project sought to promote measures to access existing systems of care for the dying, to develop appropriate new systems of care that improve the care of dying persons and to support their families and friends. PDIA provided funding for programs that examined various settings for death, including the home; identified barriers to needed services; and designed strategies to overcome them. PDIA also supported the evaluation of special programs in palliative and hospice care as well as efforts to integrate their principles of practices into the delivery of services to all dying persons. Improving continuity of care and stimulating ongoing quality improvement were also concerns, as was the dissemination and replication of promising models of care.

The fifth priority area includes educational programs for the public about death and dying. PDIA promoted the effective use of education in the media to increase discussion of dying, death and bereavement. It supported strategies for informing the public about available programs for providing support during the dying and grieving process. It supported teaching approaches that build upon the regular and unexpected experiences of students with the dying.

PDIA advanced strategies for addressing the needs of special populations in the area of educational programs, the sixth priority. For healthcare professionals, the Project supported the development of innovative curricular materials and educational methods, patient-centered clinical teaching and system level quality improvement activities to all levels of healthcare professional education. The project focused on programs that enhance the interpersonal skills needed for working with dying persons, their families and friends.

In the seventh priority area—the shaping of governmental and institutional policy—PDIA's interest in this category included evaluating and changing federal, state and local government policies around dying and bereavement; the role of large healthcare delivery systems; government reimbursement issues; the development and maintenance of an appropriate work force through, among other practices, established accreditation and certification standards; the role of professional clinical practice guidelines; and the implications of healthcare reform for the care of the dying and bereaved.

PDIA has also taken on special initiatives to address the legal and economic barriers to end-of-life care as well as the need for grassroots initiatives at community levels to address barriers to end-of-life care. Numerous educational initiatives have also been supported, including the Coalition of Medical Educators to develop curriculum in end-of-life care and the First International Research Meeting in Palliative Care.

Early on, PDIA joined with other foundations to support the Institute of Medicine report entitled "Approaching Death."(2) This monograph outlines the barriers to excellent end-of-life care and makes a series of recommendations to governmental, non-governmental, educational and public groups to encourage the development of policies and procedures to address the needs of dying patients and families. PDIA supported the Alliance for Aging Research to publish a report entitled, "Seven Deadly Myths"(6), describing the myths of the high cost of care for the elderly.

To encourage broader philanthropic funding in this area, PDIA joined with a series of foundations to create Grantmakers Concerned with Care at the End of Life (GCCEL). These funders are working to expand funding coalitions in end-of-life care. GCCEL working with the Alliance in Aging Research published a report on women and end-of-life care entitled "One Final Gift."(7) This report identifies aging and end-of-life care as an important women's issue. The development of other initiatives as well as educational opportunities for nurses, social workers, and pastoral care workers is currently part of the second three-year program of PDIA.

In its 1990 monograph entitled "Cancer Pain Relief and Palliative Care" the World Health Organization (WHO) defined palliative care and recommended that governments not consider legalizing physician-assisted suicide and euthanasia until they had addressed the needs of their citizens for palliative care services.(8) Palliative care is the term used to describe an appropriate competent compassionate system of care for patients and families. It is defined as the active total care of patients with incurable disease in which quality of life is paramount and in which control of physical, psychological and existential symptoms are of major importance. Palliative care affirms life and regards dying as a natural process. It is patient and family centered. There are five precepts of palliative care: respecting patient's goals, preferences and choices; comprehensive caring; utilizing the strength of interdisciplinary resources; acknowledging and addressing caregiver concerns; and building systems and mechanisms of support.

Both in the United States as well as internationally, there is increasing attention to the need for palliative care services to improve the care of the dying. Both developed and developing countries are beginning to focus on this important medical and social need. In both Britain and Australia, palliative care has become a medical specialty and, in fact, there are chairs of palliative care in major medical centers in both countries. PDIA supports the efforts of the WHO to make palliative care a priority.

The medical, legal and ethical debate over physician-assisted suicide and euthanasia both in the United States and internationally has been a focal point of much of the media coverage on the care of the dying.(9,10) The PDIA has tried to move the debate by calling attention to the larger issues that the movement for physician-assisted suicide reflects and conceals. Using this media attention to encourage a broad public discourse that reflects the real concerns of patients and families and encourages "death talk" in the form of asking important questions about: How will I die? Where will I die? Who will care for me as I die? Who will pay for my care? Will my cultural, spiritual and religious beliefs be respected?

In June 1997, the United States Supreme Court rejected arguments for a right to physician-assisted suicide and in essence, strongly supported the right of dying patients for palliative care.(11) Moreover, the court clearly distinguished aggressive treatment of symptoms in dying patients that might secondarily hasten their death from physician-assisted suicide and euthanasia, strongly endorsing one of the major tenets of palliative care.

PDIA serves as a model program for non-governmental philanthropic organizations that can help to create a momentum for change in transforming the culture of death. There is a need for international groups to address these important topics on the care of the dying and to see such programs as integral to programs in successful aging. From numerous public surveys, nationally and internationally, there is evidence to suggest that people have complicated and, at times, contradictory attitudes toward death and dying. They consider their own death with a mixture of dread and procrastination. They fear overuse of medical technology, want personal control and are afraid of being a burden. These attitudes can only be addressed by developing broad-based programs that address the issues of improving communication and decision-making at the end of life, changing the culture of health care institutions and changing the culture and attitude toward death. As it has been widely discussed in this conference, the growing aging population needs to participate in discussions that will facilitate their ability to obtain appropriate, competent, compassionate care at the end of life. As part of any successful aging program, addressing these issues offers the opportunity for patients to be well informed about their options of care and most importantly, to be able to advocate for care systems that respect their uniqueness, their aging and their dignity.

REFERENCES

1. The SUPPORT Principal Investigators. A controlled trial to improve care for serious ill hospitalized patients, JAMA 1995; 274:1591-1598.

2. Institute of Medicine, Committee on Care at the End of Life. Approaching death: improving care at the end of life. Washington, D.C.: National Academy Press, 1997.

3. Meier, DE, Morrison, RS, Cassel, CK. Improving palliative care, Ann Intern Med 1997; 127(3):225.

4. Project on Death in America. Report of activities: July 1994-December 1997. New York: Open Society Institute, 1998.

5. Billings, JA, Block S. Palliative care in undergraduate medical education, JAMA 1997; 278:733-738.

6. Alliance for Aging Research. Seven Deadly Myths. Washington, DC: Alliance for Aging Research, 1998.

7. Alliance for Aging Research. One Final Gift. Washington, DC: Alliance for Aging Research, 1998.

8. World Health Organization, Cancer Pain Relief and Palliative Care. Geneva: World Health Organization 1990.

9. Angell, M. The supreme court and physician-assisted suicide: the ultimate right. N Engl J Med.; 1997; 336:50-53.

10. Foley, KM. Competent care for the dying instead of physician-assisted suicide, N Engl J Med. 1997; 336:54-58.

11. Burt, RA. The supreme court speaks: not assisted suicide but a constitutional right to palliative care, JAMA 1997; 337:1234-1236.

Social and Cultural Challenges of Longevity

1. Longevity: Generational Conflict or Interaction?

INTRODUCTION

Gordon J. Piller

Abbotswold, Great Wilford
Shipston on Stour
Warwickshire
CV 36 5NQ
United Kingdom

Historically, in most societies the aged were respected and honored. It is only recently that the authority vested in older people has begun to evaporate. We now expect to see younger politicians, younger leaders of industry, and younger decision-makers across whole fields of enterprise. Youth is venerated in many developed countries. Conflict between the generations was therefore inevitable. But is it a phenomenon that will continue unchanged? The conflicts, and the areas in which they operate have changed. The older members of society are finding new roles. Their lives in developed countries take different directions in old age. Health and well-being are the key, together with the maintenance of active life. But we should not concentrate only on the wealthier societies. Our concern is also for the underdeveloped world. Their conflicts are different: they include poverty, lack of basic necessities of life, low levels of education, etc. We must address these differences when we consider generational conflicts.

I shall end with two points. The Beatles hit song "Will you still love me when I'm 64" was written in 1967 with the chorus: *Will you still need me, will you still feed me when I'm 64?* Perhaps it needs an update to: *When I'm 94*!

Maybe this is answered now by the words of Professor John Grimley-Evans, Professor of Clinical Gerontology at the University of Oxford in delivering the prestigious Harveian Oration of 1997 entitled, *A correct compassion—the medical response to an ageing society*. "The aging of populations is not a transient epidemic that will pass—it is a permanent change in the structure of society to

Longevity and Quality of Life, Edited by Butler and Jasmin.
Kluwer Academic / Plenum Publishers, New York, 2000.

which society and individuals must adapt. The new political configuration in Europe offers us wider opportunities than we have ever enjoyed for research into the intrinsic and extrinsic causes and mechanisms of aging and into the design and funding of health and social services.

"The chief impediment is likely to be political paralysis. The need for a long term view, the possibility of being confronted with difficult decisions including the issue of inter-generational inequities in funding, make it unlikely that governments will provide the leadership and vision necessary to prepare the nations for the challenges to come. The leadership will have to come from within the ranks of the knowledgeable and socially responsible citizenry."

In other words: the responsibility lies with us all if we wish to be considered "citizens" in the true sense.

THE AGE REVOLUTION: BENEFITS AND CONCERNS

James E. Birren

UCLA Center on Aging
10945 Le Conte Avenue
Suite 3119
Los Angeles, CA 90095-6980
USA

INTRODUCTION

The demographic changes in the age structure of developed societies in this century are without precedent in history. It is not surprising that there is much uncertainty in anticipating appropriate adaptations to the changing size of the different age segments. There is not only uncertainty in developed countries, who already face dramatic shifts in the age of their populations but also in developing countries who face rapid age changes in the next century. Uncertainty exists at all levels of societies, institutions, as well as in individuals themselves. This paper will review the background of the changes. It will present some options to reduce intergenerational conflict and to use the growing populations of older persons for the benefit of society, community, and family.

MAGNITUDE OF THE AGE REVOLUTION

The term "age revolution" has been chosen to describe the shift in age structure of developed societies. A revolution implies a rapid radical change in the structure of a society that requires a basic reorientation and reorganization. It is anticipated that the "age revolution" will have vast repercussions.

More years have been added to the average life expectancy in this century than were added from Roman days up to 1900. For the United States this century has seen life expectancy increase from 47 years to about 77 years. Not only have childhood years been added, but years have been added in late life as well. For white men at age 85, the average number of years remaining has increased since 1900 from 3.8 years to 5.3 years. For white women, the comparable figures are 4.1 and 6.5 years (Hobbs and Damon 1996). At age 65, the same groups have an

Longevity and Quality of Life, Edited by Butler and Jasmin.
Kluwer Academic / Plenum Publishers, New York, 2000.

average life expectancy of about 80 and 85 years respectively. The greater life expectancy, coupled with lower fertility, has shifted the age structure. The most populous group in 1900 was that of children ages 0-9. Today, it is the least populous category. By 2030 the most populous age group will be persons over the age of 65, which in 1900 was the least populous. Thus, the age pyramid is being turned upside-down. It certainly seems justified to call the age shift we are witnessing a "revolution," when children become the smallest age group and persons over 65 become the largest segment of society (Martin & Preston 1994).

The change in average life expectancy has occurred so rapidly it cannot be the result of biological evolution. While the human genome contains our biological destiny and potential, it interacts with the environment in which we grow up and grow old. Clearly our physical and social environment modulates the expression of the human genome, otherwise we would not be experiencing such an unprecedented increase in life expectancy in one century. A reasonable point of view is that human longevity is a product of human ecology, the result of the interaction of our heredity with our physical and social environments. With this perspective it seems reasonable to expect further changes in our life expectancy and in the age structure of our societies as we gain insights through research into the ecology of human aging.

It is difficult to enumerate the many changes that have occurred in our environment which have contributed to the extension of human life in this century. They range from reduced infectious diseases resulting from clean water supplies, improved sewage and refuse disposal, screens on windows that protect against insects, to improved working conditions and reduction of child labor. Improved food preservation and elimination of seasonal nutritional deficiencies have also encouraged the growth of our young and their concomitant resistance to disease. Vaccinations and antibiotics have reduced the incidence and after-effects of many serious diseases. Teaching the public sound health practices and an increased adherence to healthy life styles have also been beneficial. Our genetic inheritance is obviously being expressed in different environments from those in which our grandparents grew up and grew old. While negative environmental changes may occur, e.g., nuclear radiation and urban air pollution, the trend appears to be toward environmental changes that will extend active human life.

Emerging benefits and challenges of the age revolution

Gender work roles

In developed societies the age revolution is occurring as the industrial age is phasing out and being replaced by the information age (Vercruyssen, et al 1996). While these are not wholly independent of each other, their consequences for society are different. The industrial age brought with it the availability of large sources of energy to replace animal and manpower. For example, in the early days of farming a man might cut a half-acre of grain in a day using a hand sickle. The advent of the scythe increased this to an acre a day, and was in turn replaced by the tractor, which increased the total to a hundred acres per day per operator. Increasingly in agriculture, manpower was replaced by machine power in plowing, planting, harvesting, and in the case of grains, threshing. Large numbers of workers were removed from the agricultural labor force. The agricultural work force in the

USA has declined in both absolute and relative numbers since 1900, when it was the major work force, employing about 37 percent of the total labor force in the nation. That number is now between 2 and 3 percent, reduced from about 11 million workers to less than 3 million.

The use of high energy sources also changed the nature of skills required to do the job. Early agriculture depended heavily upon experience and acquired skills. Increasingly, education became necessary as manpower-replacing devices and energy sources were used. Strength was an advantage for the early agricultural worker in tasks ranging from plowing to animal husbandry. The strength of animals and men, horsepower and manpower, is now of less importance than the management of machines.

In industry also there has been a decided shift from the use of manpower as a source of energy, to human control over energy and the production of goods. The relative strength of a worker is now of little importance, and the control of energy and production is mediated by computer systems. It makes little difference whether a male or female finger controls the information system that releases the energy and manages the machines of production. One of the latent issues in the competition between men and women in the work place has been the redundancy of men's greater physical strength in the information age. The shifting roles of men and women are symbolic of past history, when strength was a utility, to a time when skills in the management of energy are critical. Thus, employment prospects have opened for women while concurrently men have lost functions which in past centuries provided a sense of pride. As developed societies move further into the information age occupations that prized strength and height will disappear.

Length of the work life

We are working less and living longer. Sustaining the same standard of living during long years of retirement requires a larger capital investment than anticipated in the past when age 65 was the accepted age of retirement in America (Cutler 1996). The capital for retirement is dependent upon individual savings, contributions from employers, and taxes. Since life expectancy has been steadily increasing over this century, a longer period for retirement income is needed. These sources will have to be increasingly examined for their proportionate contribution. For example, 50 million Americans are without private pensions. Since saving for retirement has not been a strong feature of the American way of life, more dependence is placed on Social Security for economic well being in retirement.

The challenge is to adjust our traditional ways of providing for retirement. This touches upon deep-seated attitudes toward taxation, savings, and providing capital for business, and it will not be a simple political process to adjust to our prospects for living longer in the information age. Since we are living longer and are healthier it may reasonably be expected that we should work longer before we retire. In addition to raising the age of retirement we will likely have to increase our personal savings for retirement. This requires that we reduce consumer spending, which has been well established in the credit card age. Small employers should be encouraged or required to set up pension plans. Since taxation is ultimately required to manage the Social Security system, adjustment of the tax level is a sensitive political issue that is being debated. Appropriate adjustments in all three sources of capital for supporting our longer retirement years touch deeply rooted social attitudes, similar in ways to those of the emergence of gender equality.

The age revolution is affecting almost all societal functions, and the institutions that provide them. The benefits and challenges must be anticipated if society is to be well-served in the next century.

Churches

The populations served by churches, synagogues and temples are growing older. Since most religious organizations give priority to the young in creating schools, teenage activities and programs for unmarried young adults, little structure is created for the elderly. In the future, religious institutions will be increasingly serving two broad populations of the elderly. They are the old who require assistance in their lives and the old who seek new personal growth opportunities. Some of this population seeks to explore its spiritual life in ways different from the young who are being given their initial orientation in faith and rituals. Seminaries that train religious leaders will need to include pastoral training oriented toward serving the older population. The next generation of religious leaders will have to expand their grasp of the religious journeys of the elderly and, arrange for services for the older congregant. (McFadden 1996).

The elderly become widows and widowers and live alone. They need social contacts within a familiar supportive environment. Long-term church membership can provide the basis for introducing a wide range of activities designed to raise the quality of life for elderly people who life alone. Volunteer work to help the needy elderly is one of many activities that can be organized. Others include afternoon adult education classes, e.g., in computer skills to establish e-mail contact with children living in remote cities, autobiography, and local oral history discussion groups. Some movements have already been started to explore this relatively new territory of religion-based service.

Education

Educational institutions as well, have traditionally been oriented toward serving the young. It has been said that universities have followed an "inoculation" concept of education in which a young person traditionally receives a large dose of knowledge which is expected to last for the rest of that person's life. Relatively few grey haired students are seen in institutions of higher learning.

The advent of the information age has been accompanied by a rapid obsolescence of jobs. Professionals as well as laborers have to update their skills. Generally, this education is left to business and industry, since less that half of worker retraining is carried out by institutions of higher learning. Thus, training in business and industry is geared toward specific skills or job knowledge requirements with little broader education to help people attain a productive and high quality of life in a rapidly changing society. Beginning in middle age, reintroduction to higher education is desirable to update the knowledge and perspectives of persons living both in the age revolution and the information age. While earlier in this century a four-year college education may have been enough to last a lifetime, the explosion of knowledge with which we are now faced requires more than even an active reader can cope with. "Booster" education shots are needed, not only for employment, but also to help individuals adapt to a changing society (Manheimer 1996). Higher education should be used to keep the population abreast of change and discovery. While it is serving the young, it also needs to react to the needs of the population for higher education across the life span.

Entertainment

Entertainment is lax in its response to the age revolution. Television and movies are for the most part oriented toward young adults. A survey of the television and movie industry indicates that it admits to being one of the most ageist of businesses. Motivated by advertising interests, the themes selected for development are geared to young adults. Advertisers have a relative blind spot when it comes to defining the mature audience as a potential market. They pick the audience to serve a business interest, a young writer (below age 40) is chosen to write the script, young directors and young actors complete the production net. As a result, the older audience receives leftover entertainment themes. A symposium on the media and the mature audience was sponsored by the Academy of Television Arts and Sciences, the American Association of Retired Persons, and the University of California at Los Angeles (AARP 1993). A variety of data sources were presented, including data on the ages of the program writers. For example, only 11 percent of the writers of episodic dramas on television networks were over age 50. Older viewers spend more hours watching television yet program themes are selected and produced by young persons. Little in the way of age-relevant themes reach the mature audience. Both the producers of entertainment and the business interests being advertized miss the fact that a potential older audience exists with the means and desire to pursue their interests for sophisticated mature entertainment. The executive director of the AARP, Horace Deets said that, "Television is reflecting a life course for older Americans that is at least 40 years out of date, if it ever has been in date." (AARP 1993, Deets, p. 41).

Individual models of aging

Not only are society's models of old age and aging out-of-date; our individual role models also need updating. Today's mature population cannot look to the role models of their grandparents and parents to guide their retirement lives. Grandparents rarely retired. Social security and pensions are now common topics for discussion but this was not the case earlier in this century.

Living longer and more actively results in people having more years to spend in pursuits of their own choosing. Birren and Feldman (1997) described the process of taking a new perspective on the ways that time is spent in the later years as rebalancing one's "life portfolio". The life portfolio was defined as the profile of time devoted to one's major interests, activities, and concerns. Reexamining the ways we spend our time can help us to best utilize the gift of long life this century has given us.

Using the benefits of the age revolution

The dramatic growth of the mature population brings with it a wealth of experience that can be used to guide the family and society as a whole. Greater attention should be given to utilizing this experience in the public sector. For example, retired heads of government agencies might be convened at regular intervals to present their views on emerging issues. Mature persons have lived though periods of inflation, prosperity, economic depression, war, and problems relating to the natural environment. This experience should be valued, and while not necessarily used to dictate the course of every action taken, at least considered. "Senior councils" might be developed comprised of former leaders and those who have lived long lives in changing environments. These councils could be created across political, philosophical and cultural lines and serve to utilize the wealth of experience of the retired.

CONCLUSION

The age revolution of this century brings with it many challenges for societies, institutions and individuals. Our models of old age are out-of-date yet many planning efforts ignore the fact that older persons are replacing children as the most populous segment of society. In developed societies, retired persons are not only living longer; they are healthier, more active, and better educated. Increasingly, planning efforts will need to take into account the trends described in the research literature about the size and nature of the new older populations. Individuals will also have to update their concepts of aging as they live long years—years that can be productive both for them and for the societies in which they live. Research on the processes of aging should be encouraged at all levels–social, behavioral and biological–so that we can gain the knowledge to enable us to live fuller, healthier and more productive lives and to further extend the gift of life.

REFERENCES

1. American Association of Retired Persons. 1993. Age has a future: maturity and the media. Washington, D.C.

2. Birren, J. E., & Feldman, L. 1997. Where to go from here. New York: Simon & Schuster.

3. Cutler, N. E. 1996. Pensions. In, J. E. Birren, (ed.), Encyclopedia of gerontology, pp. 261-269. San Diego, CA: Academic Press.

4. Hobbs, F. B., & Damon, B. L. 1996. 65+ in the United States. Special studies, P23-190.Washington, D.C.: Bureau of the Census.

5. Manheimer, R. J. 1996. Adult education. In, J. E. Birren (ed.), Encyclopedia of gerontology. Pp. 61-69. San Diego, CA: Academic Press.

6. Martin, L. G., & Preston, S. H. (eds.). 1994. Demography of aging. Washington, D.C.: National Research Council.

7. McFadden, S. H. 1996. Religion, spirituality, and aging. In, J. E. Birren, & K. W. Schaie (eds.). Handbook of the Psychology of Aging. San Diego, CA: Academic Press.

8. Vercruyssen, M., Graafmans, J. A. M., Fozard, J. L. Bouma, H. Rietsema. 1996. Gerotechnology. In, J. E. Birren (ed.). Encyclopedia of gerontology. San Diego, CA: Academic Press.

ARE WE MOVING TOWARD A WAR BETWEEN THE GENERATIONS?

Claudine Attias-Donfut

Caisse Nationale D'Assurance de Viellesse
49 rue Mirabeau
Paris 75016
France

INTRODUCTION

Is there a generational war threatening aging societies? If so, it will not be a cultural conflict, as in the 1960s, arising from young people opposing gender and generational domination, but rather an economic conflict. The success of the 60s movement has brought about changes in generational relations: they have become closer, less hierarchical and more cooperative. Moreover, the rise in the numbers of older people which has occurred as a result of increased longevity, and hence their greater proportion in the population has also completely transformed the relations between generations. People are living longer and the overlap in the generations is increasing. It is no longer unusual to reach the age of retirement while still having parents alive, and therefore to become an "orphan with white hair." What price are we paying for such changes, and who is paying for them? Can successive generations support a period of retirement which lasts longer and is more costly? The times have certainly changed since Kant wrote "previous generations always seem to have given everything for the benefit of generations to come. . ..."(1) New values which emphasize quality of life, together with living for the present moment, render it inconceivable that one generation should make large sacrifices for the benefit of another, whether this be for the generation to come or the previous one.

The contemporary importance placed on the self and individual autonomy explains why the theme of "generational equity" is in ascendance, a theme which would have seemed absurd to someone in the eighteenth century. Such an idea is indicative of the philosophical questions relevant to post-modern societies, bringing into question the whole notion of time and historic fact. The problem of justice between generations has been analyzed by John Rawls in the context of his theory. According to Rawls, generations are linked by an implicit contract, each generation transmitting to the following generation the equivalent of what it had received from the preceding one. Therefore, to achieve justice, a "fair principle" of savings is required by each generation.

Longevity and Quality of Life, Edited by Butler and Jasmin,
Kluwer Academic / Plenum Publishers, New York, 2000.

Following Rawl's theory of justice this problem has been the subject of many studies, primarily among economists, and has given rise to lively debates in the USA and in Europe. (e.g., Bengtson & Achenbaum 1993; Laslett & Pishkin 1992; Attias-Donfut 1995; Masson 1995; Walker 1996).

But generational equity is also a deceptive notion open to manipulation when applied to the sharing of work and measures of social protection within a society. Systems of social protection are precisely what are at stake in the so-called war between generations. The argument is well-known: increased expenditure on health and pensions for older generations will occur as a result of greater life expectancy. This expenditure will fall more heavily on the working population, hindering the development of job opportunities and contributing to unemployment and poverty among younger age groups. The resulting inequality between generations will be the origin of the conflict, setting the young and the active against the retired (Sgritta 1997). It is often said by those working within this framework that inequalities between the generations will replace social inequalities, or similarly, that the importance of older people in terms of political votes will be an obstacle to any necessary reforms. This discourse also links numerous other problems of unemployment, social inequalities and social protection, to political and ideological debates. These problems are fuelled by a pessimism arising from the decline in the growth of welfare states. There has been somewhat of a time gap between the improvement in standards of living among the retired population and the period of prosperity in the post-war period (the "Thirty Glorious Years"). Moreover, these improvements have taken place at the same time as the growth of economics began slowing and young people began to experience difficulties in securing employment. Today a new situation has arisen, where young adults run the risk of falling into social class groups lower than those of their parents. Young people therefore often receive some form of help from their parents and grandparents.

There can be no doubt that, from an ethical or philosophical perspective, the concept of generational equity is an important one. But it can also take on quite a different meaning insofar as it concerns the Welfare State. In these cases, debates on generational equity mask, in fact, the old debates between conservatives and democrats about the Welfare State. These debates are mainly held among experts and have no bearing on what is happening in the population.

The following three points, to be developed in detail, clearly demonstrate this assertion:

- opinion polls taken in many nations do not indicate generational tensions around the issue of retirement pensions,
- private transfers exist between generations in the family, which complement public transfers,
- public and private transfers between generations reduce intergenerational inequalities. Consequently, public forms of solidarity do not increase the risk of a war between the generations but, on the contrary, reinforce existing ties.

AN INTERGENERATIONAL CONSENSUS ON PENSIONS

Opinion polls in various European countries confirm the public's preference, among older age groups as well as younger ones, for the value of retirement pensions to be equivalent or similar to salary levels. Two separate Eurobarometer studies illustrate this preference clearly, as has been shown by Kohli (1995).

The first study, undertaken in the Spring of 1992, shows that four-fifths of the adult population of the European Community believe that a person in paid employment has an obligation to guarantee a decent standard of living to older people through the payment of taxes. Fifty-seven per cent believe that the level of resources that the State should offers older people be a sum approximate to the average wage. Very few people believe that the figure should be lower. The second study, undertaken in several European countries between 1985 and 1990, also shows a very strong attachment to maintaining adequate pension levels. The majority of people interviewed were in favor of increased public expenditure to secure adequate pensions, even if such a measure necessitated extra taxes. Even in the United States, where many children and young people live in poverty and the debate on generational equity is very topical, the majority of the adult population (including those aged between 18 and 25 years), are in favor of more public spending for pensions.

Finally, the study showed that over a five-year period, the proportion of the population in favor of strong intergenerational links grew. Among the youngest and oldest generations, increasing the levels of pensions was not shown to give rise to polemic debates.

Such measures are viewed as security for the future and as a counterpart of the contributions made during a person's working life, conforming to a moral contract between the individual and society. The debate on generational equity therefore does not take place for the public in the way that it does for academics and specialists. Those who challenge generational inequity are more concerned about an overall reduction in social benefits, rather than a redistribution of resources away from older to younger generations. The discourse on generational equity is an ideological ploy to promote private forms of social security by setting up the generations in conflict. It is a discourse that extends widely to the media in western countries, which feeds upon the anxiety that most people have over their retirement. In fact, the controversies over the future of retirement pensions is projected onto an imaginary conflict between the generations.

GENERATIONAL ACCOUNTING, AN INCOMPLETE SYSTEM

The debate on generational equity is mostly restricted to the question of pensions to the extent that it ignores financial transfers within families, which complement public transfers. These family exchanges have been measured in research on three-generational families, where a member of the intermediate generation (aged between 49 and 53 years), a parent and a child were interviewed. This research measured intensively the services exchanged between the three generations in everyday life, including such areas as domestic or social activities, living arrangements and personal care, as well as regular and occasional transfers of money, inheritance and gifts.(2)

These three generations in the research represent relatively homogenous age groups, clearly differentiated with respect to historical periods and the stages of their life course; the young are in the process of entering the world of work (jobs for some, further education for others), the "pivot" (intermediate) generation is for the most part active, and the eldest generation has been in retirement for some time and is at risk for experiencing disability.

Within this context the different definitions of "family", "historical" and "welfare" generations coincide, and by constructing the sample in this way it is rendered particularly suitable to an analysis of the interface between private and public intergenerational transfers. Our results, as they relate to the different standards of living relative to the generations, show inequalities of income between different age groups, but at the same time they show how the current resources of the pivot and younger generations are more substantial than among the eldest generation. This finding is particularly significant for older widows, and men and women over the age of 75 years. While improvements in the standard of living have above all taken place for the recently retired group, the oldest of the old and widows continue to have low incomes.

The age of younger people also influences the availability of the most important resources, with the majority of respondents aged between 20 and 32 years. Many of these younger adults are employed. The situation of working younger adults differs substantially from those who are finishing their education and facing the current difficulties of entering the labor market. For these younger adults, transfers received from the older generations are often their sole source of income. In this sense, these young people are not a homogenous group representing a "welfare generation", defined according to whether they are active in the labor market and depending upon their status regarding contributions to, or redistribution of the social security budget. This evidence from our research obscures slightly the opposition between availability of resources and the generations as well as further emphasizing the existence of strong intergenerational differences.

An analysis of the composition of resources within families confirms the theory that public transfers are principally for the benefit of older generations. The resources of older age groups are made up mostly from their pensions, which are determined by the social contributions of the active younger and pivot generations. These contributions, mainly in the form of taxes, were limited since they are proportional to income. On the other hand, these payments accounted for only one-tenth of the total resources of the active generations (Attias-Donfut & Wolff 1997). Young people received much of their resources by way of state financial help for their education (grants and housing benefits) or benefits related to unemployment (such as unemployment benefit or income support). A certain number of pivot generation households were in receipt of retirement pensions (either through early retirement or by means of their spouse's pension) and other benefits for those who were not active in the labor market.

Contrary to the common approach of juxtaposing the generations (an approach which considers only public transfers, e.g., generational accounting, cf. Kotlikoff, 1992; Auerbach et al,1994), we also examined the flow of private transfers between the generations. The totality of these transfers is, however, difficult to calculate because they ought to take into account retrospective data relating to the life course based on inheritance wealth, gifts and other economic aid either in nature or in kind, received either habitually or occasionally. In the absence of such estimations (which in reality are difficult to make), our data examined the actual transfers made relating to gifts and loans over the past five years. The combination of responses simultaneously received from three generations of the same family made it possible to determine that, among grandparents (the older generation) 33% gave money to their adult children and 30% give money to their grandchildren (the younger generation). In total, almost one in two older people (49%) gave money to either their adult children or grandchildren. As far as the pivot generation is concerned, the proportion of donor households is 64% for financial help to children and 9% for financial help to their parents (Attias-Donfut 1995). Descending generational transfers coexist with ascending transfers which are much less limited in scope.

This analysis enables us to determine two facts. On the one hand, state financial help benefits mostly older generations, who in turn are beneficiaries of a system into which they paid when they were economically active. On the other hand, flowing from this ascending pattern of transfers, private transfers for the most part benefit descending generations. Older generations redistribute their financial resources not only to their adult children (the pivot generation) by way of gifts and inherited wealth, but equally to their grandchildren in the form of timely offers of financial help and gifts. However, in contrast to public transfers which are regularly bestowed, the total amount of private support is unknown and not all households display this type of solidarity. Besides, the informal and sporadic character of support makes it impossible to establish the exact amounts which would allow a precise comparison between public and private support. Our analysis, however, does show the existence of circular mechanisms in the pathways of support, whereby public support largely supplement and add to the private support systems between the generations.

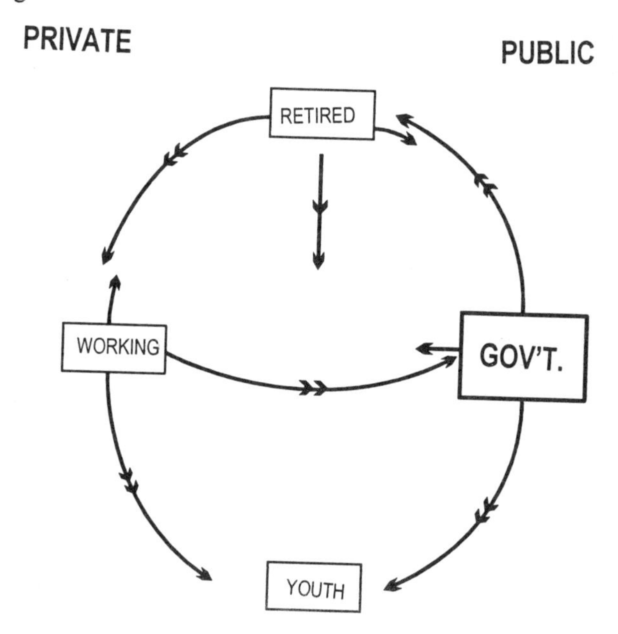

Figure 1 : Exchanges between the three generations

These results bring into question the idea of generational conflict, a notion which to a large extent originates from considering only public transfers (the right half of Figure 1). The cycle of exchange shows the interdependence of the generations, and the sustaining elements of this reciprocal process between private and public support. With the provision of pensions, the State has put into place a new structure of family solidarity. Whereas in pre-industrial societies younger generations took responsibility for aging parents, today it is parents and grandparents who support and financially help their descendants. This marked bias toward descendants which is a feature of modern society is the product of an evolutionary process(3) and should not be considered as a universal fact which is historically common to all kinship relations.

These circular mechanisms, with their descending flow of familial distribution, are only possible today because the "very old" enjoy a quality of life that was previously unknown. Not only can they remain independent without having to rely upon financial help from their children, but they can also generously provide for their descendants. This "life span" aspect of retired people provides a strong case for the development of pensions in the public sector (Caussat 1996).(4) Moreover, since the risk of being without resources in advanced age has largely disappeared, the inheritance wealth of older generations is not suddenly cut off. Rather, this wealth continues to grow in a period which has seen an expansion of the property market and low inflation.(5) Finally, the importance of adequate social policies which benefit older people should be emphasized, since a significant proportion of retired people contribute toward and supplement the family economy.

A REDUCTION IN INEQUALITIES BETWEEN THE GENERATIONS

The greatest inequalities today are those encountered in the labor market. Social benefits clearly have a role in reducing these inequalities (Piketty, 1997). Moreover, our research has shown that intra-family support has the largest effect upon the most disadvantaged members, as for example the youngest and oldest generations. For young adults, the most important sources of economic help are directed toward helping them achieve the transition to adult life (by helping them to find paid work and setting up a family) since this is an uncertain time of life. Whereas young adults, who have the smallest incomes, receive more regular forms of support (especially students, but also the unemployed), parents who are well-off are the most regular donors of financial support. However, parents with modest incomes give proportionately twice as much of their income to help their children compared with the most affluent parents.

Family support and professional services differ according to social class. In higher social classes, older people in need of support can turn to the private sector, while public support tends to be targeted toward the lower social classes (Renaut and Rozenkier 1995). This public support organized by the State is means-tested, and directed toward very old people, who have lower incomes than younger retired people. The very old who receive public support are mostly widows with very low incomes (even though their incomes were improved during the period 1970-1984).(6)

Family support also tends to be targeted toward the poorest elderly in the same way. Such support is given mostly by children and it takes several different forms. These can be cohabitation, financial gifts, and practical forms of support. Each one of these forms tends to decrease in volume when the incomes of older parents are high. Family support results in the improvement of standards of living for the poorest and most vulnerable sectors of the older population, and therefore contributes toward alleviating inequalities.

It is difficult, however, to evaluate the full impact of this process, especially when it consists of caring activities or cohabitation. When only financial resources are taken into account, the total amount of family support is largely underestimated.

Furthermore, financial gifts from older people to their descendants reduces the availability of their own resources. These different types of intergenerational transfers are shown in Table 1.

Table 1: Private transfers according to the income of the elderly

Income of the elderly	Quartile 1	Quartile 2	Quartile 3	Quartile	Total
Type of support (Frequency)					
Cohabitation with children	29.3	17.1	10.2	8.2	16.2
Financial help from children	3.9	4.9	3.6	1	3.4
Daily domestic help from children	50.3	42.8	33.2	21.6	37
Financial help to children	12.5	24	24	41.6	25.6

Source: CNAV Trois Générations Survey 1992.

Note: For each of the different types of support, the figures reported concern the proportions of help given.

This circular flow, upward from the pivot generation toward the poorest older generations and downward from the richest older generations to the pivot and younger generations, has contradictory effects upon the gap in standards of living for the older generations. The redistribution of resources within the family substantially increases the standard of living of the younger generation, whether they are inactive or at the beginning of their careers. The results of the Budget Survey of Families (Barry et al 1996) also confirm our findings. Financial transfers within families significantly increase the incomes of household members aged under 40 years, and above all they increase the incomes of young households. For households where the members are aged above 40 years, the transfers they make reduces their disposable incomes. The net effect of these transfers is to reduce inequalities of incomes that are age related.

The closer level of standards of living between the generations that is due to the circulation of financial transfers between them can also be seen in the forms of support that are given to the oldest generations. This support can be financial, cohabitation or can take the form of providing care services in the case of disability. These forms of help are particularly beneficial to the most vulnerable of the older generation and they are mostly provided by children who have experienced upward social mobility. This process therefore represents a reverse effect of social mobility patterns.

Social benefits help mechanisms described above, since they in turn facilitate either an intra or intergenerational sharing of resources that are directed to the poorest groups. This process, which on the one hand can be empirically determined at the macrosocial level, is generally applicable to the wider population, although there are certain sectors, such as those without family members (or those with dysfunctional family relations) who are excluded. On the other hand, whereas support from within the family can also help to limit temporary situations of poverty, it cannot redress more permanent situations which transcend the generations and affect families who do not have the means to engage in such transfers.

CONCLUSION: PUBLIC SOLIDARITY IS THE CREATOR OF BONDS BETWEEN GENERATIONS

These results, as they relate to French society, can be more generally applied to European countries. Comparative research shows that poverty among children and young people on the one hand, and poverty among elderly people on the other hand, go together. The more developed the welfare system, the less inequalities between and within generations. Retirement pensions and other social benefits do not lead to generational warfare, but on the contrary, produce a social bond between generations.

Improvements in the level of retirement pensions have had the double effect of allowing older people to become economically independent so that they no longer (except in a minority of cases) have to depend upon their children, and of making possible an economic contribution to the flow of exchanges within families. In this way, retired people can combine their efforts with those of their children to help the youngest generations with the difficulties that they face in establishing adult life. Capital that is saved thanks to retirement pensions allows gifts of money to be made: many young people are able to follow their studies while being supported by both their parents and their grandparents. This help also applies to many young people who have difficulties in finding employment. Moreover, adults who are active in the labor market are more likely to willingly help their children if they can be sure of receiving a good pension which is financed collectively from within a welfare system. Since they are not as concerned with having to ensure that they will have an adequate income in old age, they can focus their efforts upon the new generations. If family support systems have been able to withstand the contemporary economic crisis, it is because of the existence of public supports We can therefore draw some conclusions concerning the consequences of a move backward in social protection, in support of noncompulsory private insurance and pension funds operating on the principle of individual savings (where those with higher incomes have a specific advantage).

In addition to the evident inequalities that would follow, such systems force individuals to save for their retirement and therefore to designate less to helping with the education of their children. As such they represent a real threat to the social ties that compulsory systems reinforce or even produce. International comparisons uphold this hypothesis, and one can see that those countries where intergenerational systems of support are put into practice also have the strongest welfare state systems. Intergenerational relations have undergone a radical transformation with the modernization of society, and as social policies change they are certainly going to evolve even more. Without the support of strong social policies, the solidarity between generations runs the risk of becoming exhausted as a result of the heavy demands that are placed upon the family by increased longevity and the rise in the numbers of "very old" people on the one hand, and the growing needs of young people in the face of unemployment on the other hand. The same forces which threaten systems of social protection will eventually weigh upon the family. Far from functioning as communicating vessels whereby a reduction of support in one sphere leads to the rise in the other, public forms of support reinforce and uphold private forms of support while serving to buffer social inequalities. When social policies are reviewed, it is vital to take into account the consequences for the functioning of intergenerational support systems and more widely, the consequences upon social ties in general.

REFERENCES

1. Arber, S. & Attias-Donfut, C., eds. 1999. The Myth of Intergenerational Conflict, Family and State in Aging Society. Routledge (forthcoming).

2. Attias-Donfut, C. 1995. Le double circuit des transmissions. In: Les solidarités entre generations: Vieillesse, Familles, Etat., Nathan, Coll. “Essais & Recherche,” Série “Sciences Sociales,” pp. 41 - 81.

3. Attias-Donfut, C., Lapierre, N. 1997. La famille Providence: trois générations en Guadeloupe. La Documentation Française.

4. Attias-Donfut, C., Wolff, F.C. 1997. Transferts publics et privés entre générations: incidences sur les inégalités sociales, Retraite et Société 20: 20 - 39.

5. Auerbach, et. al., 1994 “Generational Accounting: A Meaningful Way to Evaluate Fiscal Policy,” Journal of Economic Perspectives 8, 1: 73-94.

6. Bengtson, V.l.; Achenbaum W.A., eds. 1993. The Changing contract across generations. Aldine de Gruyter.

7. Caussat, L. 1996. Retraite et assurance contre l’incertitude de la durée de vie. Retraite et Société 14: 95 - 111.

8. De Barry, C., Eneau, D., Hourriez, J.M., 1996 “Les aides financières entre ménages,” Insee Premières, 441, avril.

9. Hourriez J. M.; Legris B. 1995 “Le niveau de vie relatif des personnes âgées,” Economie et Statistique, 283 - 284, pp. 137 - 158.

10. Kohli, M. 1995 “La présence de l’histoire,” in C. Attias-Donfut ed., Les solidarités entre génération. Vieillesse, Famille, Etat, Paris, Nathan, Coll. “Essais & Recherche,” Série “Sciences Sociales,” pp. 245 - 258.

11. Kotlikoff, L. J. 1992 Generational Accounting - Knowing Who Pays, and When, for What we Spend, Free Press, New York.

12. Laslett, P. Fishkin J.S., Eds 1992, Justice between Age Groups and Generations, New Haven, Yale University Press.

13. Legris, B., 1996 “Revenus et ressources des retraités. Le niveau de vie relatif des retraités: les situations contrastées entre les très vieux et les veuves d’une part, et les retraités bénéficiant de bons revenus d’autre part”, Retraite et Société, 14, pp.25 - 49.

14. Marpsat, M., 1991 “Les échanges au sein de la famille,” Economie et statistique, 239, pp. 59 - 66.

15. Masson, A., 1995 “L’héritage au sein des transferts entre générations: théorie, constat, perspectives,” in C. Attias-Donfut ed., Les solidarités entre générations. Vieillesse, Famille, Etat, Paris, Nathan, Coll. “Essais & Recherche,” Série “Sciences Sociales,” pp. 279 - 325.

16. Piketty, T., 1997 L’économie des inégalités, Paris, La Découverte.

17. Renaut S., Rozenkier A., 1995. Les familles à l’épreuve de la dépendance in C. Attias-Donfut ed., Les solidarités entre génération. Vieillesse, Famille, Etat, Paris, Nathan, Coll. “Essais & Recherche,” Série Sciences Sociales. 181 - 208.

18. Rawls, J., 1971. A theory of justice. Cambridge, Harvard University Press (traduction française; Théorie de la justice - 1987 - Paris, Edition du Seuil.)

19. Sgritta, G. B., 1997 "Solidarité étatique versus solidarité familiale. La question des générations," in J. Commaille, F. de Singly, eds., La question familiale en Europe, Paris, L'Harmattan, Coll. "Logiques Sociales," pp. 201 - 220.

20. Walker, A., Ed, 1996, The New Generational Contract - Intergenerational relations, old age and welfare. London: UCL Press.

A REMARKABLE LACK OF INTERGENERATIONAL CONFLICT: HOW SHOULD GOVERNMENT SPENDING BE DIVIDED BETWEEN YOUNG AND OLD

Humphrey Taylor

Louis Harris and Associates, Inc.
111 Fifth Avenue
New York, NY 10003
USA

Large majorities of all age groups believe government spending balance is "about right."
Very few people, in all age groups, favor cutting programs for children or the elderly.

INTRODUCTION

Many economists and politicians see the choices which government—federal, state and local—must make as involving a clear conflict between spending on programs for the elderly and those for children. Social Security and Medicare absorb ever larger proportions of government spending and the gross domestic product (GDP). Meanwhile, the Medicaid program, which provides insurance coverage for millions of low-income children and public school education budgets are also increasing.

The experts may believe that this sets up a battle between the generations for scarce government resources. However, a new Harris Poll shows that the public totally rejects this idea. Large majorities of Americans of all ages think that when it comes to choosing between spending for the elderly and spending for children, we have got the balance about right. Very few older people want to change the balance to increase spending on the programs on which they depend. And very few younger people, or the parents of children, support changing the balance the other way in favor of relatively more spending on education or other programs for the families with children.

These are some differences between age groups, but the surprise is that they are only differences of degree, and not of direction.

Longevity and Quality of Life, Edited by Butler and Jasmin.
Kluwer Academic / Plenum Publishers, New York, 2000.

This Harris Poll is based on a nationwide telephone survey of 1,011 adults in the United States, interviewed between April 22 and 27, 1998.

MAJOR FINDINGS OF POLL

Very few people believe we are spending too much on programs for old people (6 percent), people with disabilities (6 percent), children (9 percent), poor people (15 percent) or mothers of young children (16 percent). Majorities believe we are spending too little on the first four of these, and a 47 percent plurality believe we are spending too little on the mothers of young children.

The differences between various age groups on whether we are spending too much on programs for these groups are quite modest. Few older people think we are spending too much on programs for children (9 percent) or for mothers of young children (17 percent). Few people of any age group (less than 10 percent) think we are spending too much on programs for older people. People are much more likely to believe government is spending too little on four specific programs than too much:

- for public school education, 66 percent say too little and 10 percent say too much;
- for Medicare, 53 percent too little and 8 percent too much;
- for Social Security, 48 percent too little and 10 percent too much;
- for Medicaid, 43 percent too little and 13 percent too much.

Again, the differences between the age groups in response to this question are modest. The biggest difference is in public school education where 17 percent of people over 65, but less than 10 percent of all age groups under 50, believe we are spending too much.

In the final question, where people were asked to choose between either spending more on government services for children and less for older people *or* spending more on programs for the elderly and less for children, 67 percent of all adults chose to keep the balance the same as it is now. However, three times as many people would tilt the balance in favor of more spending on programs for children (22 percent) as would increase the proportion of spending for older people (8 percent).

This question does show quite sharp differences (of degree, not direction) between different age groups, although 59 percent or more of each age group favors keeping the balance the same. Substantial minorities of younger adults (35 percent of people under 25, 27 percent of those aged between 25 and 29, and 23 percent of those aged between 30 and 39) favor increasing the share of spending on services for children. But, as the other questions show, they actually tend to favor spending more on programs for both the young and the old.

The picture which emerges from these data, therefore—whatever the experts may believe—is one of remarkable harmony and agreement between different generations, and a marked lack of intergenerational conflict. One reason for this is surely that we have multiple needs both as individuals and as families. Younger people have parents and grandparents, and expect to grow old themselves. Older people have children and grandchildren. Extended families need services for the young and the old.

METHODOLOGY

This Harris Poll was conducted by telephone within the United States between April 22 to 27, 1998 among a nationwide cross section of 1,011 adults. Figures for age, sex, race, education, number of adults and number of voice/telephone lines in the household were weighted where necessary to bring them into line with their actual proportions in the population.

In theory, with a sample of this size, one can say with 95 percent certainty that the results have a statistical precision of plus or minus 3 percentage points of what they would be if the entire adult population had been polled with complete accuracy. Unfortunately, there are several other possible sources of error in all polls or surveys that are probably more serious than theoretical calculations of sampling error. They include refusals to be interviewed (non-response), question wording and question order, interviewer bias, weighting by demographic control data and screening (e.g., for likely voters). It is difficult or impossible to quantify the errors that may result from these factors.

These statements conform to the principles of disclosure of the National Council on Public Polls.

TABLE 1

WHETHER GOVERNMENT SPENDS TOO MUCH OR TOO LITTLE ON PROGRAMS FOR POOR, CHILDREN, PEOPLE WITH DISABILITIES, OLD PEOPLE

"Many government programs cost a lot of money. Some are designed to help children, or old people or people with disabilities, or poor people and so on. In general, would you say that federal, state and local governments spend too much or too little or about the right amount on programs which provide money or services for children?"

Base: 1011

	Too Much %	**Too Little** %	**About Right** %	**Don't Know** %
Services for:				
Mothers of young children	16	47	33	4
Poor people	15	53	29	2
Children	9	64	26	1
People with disabilities	6	53	37	4
Old people	6	61	31	1

TABLE 2A

AGE DIFFERENCES: THOSE WHO BELIEVE GOVERNMENT SPENDS TOO MUCH ON PROGRAMS FOR POOR, CHILDREN, PEOPLE WITH DISABILITIES, OLD PEOPLE, ETC.

	AGE					
	18-24	**25-29**	**30-39**	**40-49**	**50-64**	**65-over**
Base	97	77	192	209	219	180
	%	%	%	%	%	%
<u>Services for</u>:						
Children	6	7	7	11	10	9
Mothers of young children	7	20	15	14	26	17
Poor People	17	15	18	14	14	13
Old people	9	5	6	5	7	6
People with disabilities	4	6	10	6	4	6

TABLE 2B

AGE DIFFERENCES: THOSE WHO BELIEVE GOVERNMENT SPENDS TOO MUCH ON PROGRAMS FOR POOR, CHILDREN, PEOPLE WITH DISABILITIES, OLD PEOPLE, ETC. (Continued)

	With Children Under 18	Without Children Under 18
Base	377	631
	%	%
Services for:		
Children	10	7
Mothers of young children	15	18
Poor People	17	14
Old people	5	7
People with disabilities	8	5

TABLE 3

WHETHER GOVERNMENT SPENDS TOO MUCH OR TOO LITTLE ON MEDICAID, SOCIAL SECURITY, PUBLIC SCHOOLS AND MEDICARE

"In general would you say government spends too much, too little, or about the right amount on . . ."

Base: 1011

	Too Much %	Too Little %	About Right %	Don't Know %
Medicaid -- the health insurance program for low-income families and children	13	43	40	4
Social Security	10	48	39	2
Public school education	10	66	24	1
Medicare -- the health insurance program for the elderly and disabled	8	53	36	2

TABLE 4

AGE DIFFERENCES: THOSE WHO BELIEVE GOVERNMENT SPENDS TOO MUCH ON MEDICAID, SOCIAL SECURITY, PUBLIC SCHOOLS AND MEDICARE

	AGE							
	18-24	**25-29**	**30-39**	**40-49**	**50-64**	**65 & Over**	**With Children Under 18**	**Without Children Under 18**
Base	97	77	192	209	219	180	377	631
	%	%	%	%	%	%	%	%
Medicare	4	10	8	9	8	7	7	9
Social Security	11	11	13	12	6	6	11	10
Public school education	3	5	9	8	12	17	6	12
Medicaid	8	19	14	12	14	10	14	12

TABLE 5

IF FORCED TO CHOOSE, SPEND MORE MONEY ON SERVICES FOR CHILDREN AND LESS FOR OLDER PEOPLE -- OR VICE VERSA

AIf you had to choose ONE of the following which would you choose? A

		AGE							
	Total	18-24	25-29	30-39	40-49	50-64	65 & Over	With Children Under 18	Without Children Under 18
	1011	97	77	192	209	219	180	377	631
	%	%	%	%	%	%	%	%	%
Spending more money on services for older people/ less on services for children	8	-	13	9	11	9	4	9	7
Spending more money on services for children/less on services for older people	22	35	27	23	19	20	16	26	19
Keeping the balance between spending for older people and children the same as it is now	67	63	59	65	67	70	75	62	72
Don't know	2	1	1	2	2	1	3	2	4
Refused	1	*	-	*	*	*	2	*	1

2. Aging and Cultural Diversity

SOME AFRICAN VIGNETTES

Kihumbu Thairu

University of Nairobi
Chiromo Campus
Nairobi
Kenya

INTRODUCTION

In a continent where life expectancy at birth is as low as 38 years in some communities it may come as a surprise to discover that a few people do live beyond the age of 90. This paper is a collection of vignettes of the lives of five such individuals.

VIGNETTE 1: THE GREAT OLD MAN OF KILIMANJARO
Johana Lauwo, Age 125

Johana was the local mountain guide who in 1989, at the age of 18, led the first European expedition to the peak of the highest mountain in Africa. For the next 60 years, he led other climbers up the mountain before retiring to his home village in the highlands near Kilimanjaro. The Tanzanian government built him a new cottage as a gift. Here he settled with his two wives to a life of small scale mixed farming. He remained in fairly good health until his death in 1996 at the age of 125. He had five daughters.

In his long life, Johana Lauwo, received numerous commendations and honors for his work. The UK awarded him the OBE for bravery in saving the life of a stricken mountaineer. He was also commended by the German government.

Longevity and Quality of Life, Edited by Butler and Jasmin.
Kluwer Academic / Plenum Publishers, New York, 2000.

VIGNETTE 2: A PIONEER IN FEMALE RIGHTS AND EDUCATION
Leah Nyangendo, Age 107

Born in Kenya in 1888, Leah Nyangendo lost her first husband in the First World War. She married her husband's younger brother as was the custom in those days. She and her second husband were the first educated Christian couple in their locality. They built the first school and church on their own land with their own resources. Their school was free, and they taught basic subjects as well as Christianity to all who wished to learn.

She turned her home into a boarding school for girls who had converted to Christianity. By virtue of rejecting female circumcision and their traditional religion, these young women had been rejected by their own families and communities. Thirty years later, female circumcision was abolished in the region. Her graduates had shown by practical example that a young woman could become a Christian, get educated, remain uncircumcised and still manage to marry, have a normal family and, despite all these changes, uphold high moral standards in her own personal life and in her family. The school that Leah Nyangendo and her husband began is now a community high school.

As they grew old, Leah and her husband became the center of their family and clan. She enjoyed good health up to the last three years of her life when she needed a house helper to look after her.

VIGNETTE 3: STILL GOING STRONG
Stephen Njoroge, Age 98

Stephen Njoroge was one of the first educated Africans in his community. He was educated by both the English and Scottish missionaries because he attended both schools, one during the day and the other in the evenings, so as to complete his studies in the shortest time possible.

After a short career as a teacher and clerk on a colonial white farmer's plantation, Mr. Njoroge settled to a life of intensive small scale farming in his own village in the central highlands of Kenya in 1930. He has not yet retired, although nowadays he only does dairy farming. He weeds all the meadows on his farm by hand.

In his younger days, Stephen Njoroge was credited with the introduction of modern agriculture to his locality through volunteer teaching and practical example. For many years he was awarded prizes by the government for having the best modern farm in his District. His farm was also used by the government for over 20 years for teaching modern farming methods.

Like other old men in his society, he is the psychological center of the life of his family. He lives independently. Since his wife died at the young age of 78, he is looked after by his daughter-in-law and her children, who live on the farm.

VIGNETTES 4 and 5: THE GATHIGIRA BROTHERS
Zedekiah Kiai , Age 102 and Stanley Kiama, Age 106

When the first missionaries arrived in Tumutumu, in the central highlands of Kenya, the father of the Gathigira brothers, Mr. Gathigira wa Muema donated land for a school. The two brothers enrolled in the same class in 1908 and each graduated with an English Certificate of Education. They then went their separate ways.

Kiai

Kiai went to Kajiado in Masailand where he worked as a teacher in the morning, and as an interpreter for the local British District Administrator in the afternoon. In 1923 he returned home to marry but was asked by his parents to first fetch his elder brother, so that Kiama could marry first, according to Agikuyu custom. Kiai married in 1925 after which he taught in a colonial European "white" school in Nairobi, before being persuaded to return home and teach his fellow Africans.

In World War II he worked as an army clerk in Kenya and later served in Ethiopia and Burma under Major Hardy. He also traveled widely, visiting Australia, the U.S., and West Africa. After the war, he became a clerk in the Tumutumu school. During the liberation struggle (Mau Mau 1952 to 1960), he was arrested by the British, and like other Agikuyu men (60% of the whole adult male population) he was incarcerated in various detention camps. After finally being released he worked as a labor officer in his hometown for 6 years and then retired to a rural life of farming in Nyahururu in the central highlands. Asked by a journalist what his secret for longevity was, Gathigira said, it was the joy of watching his children and grandchildren grow and the pleasure of spending days talking to them under a tree.

Kiama

Kiama went to work in a white settler's plantation until 1923 when his parents requested he return home and get married to enable his younger brother to marry. After marriage, Kiama became a teacher in his old school. He was later promoted to the post of Inspector of Schools and was eventually in charge of 63 schools, most of which he helped develop. He also worked as a chief (which in the central highlands was a colonial civil service creation). In 1933, he published his first book on Agikuyu culture which remains a standard reference work even today. He published his second book in 1958 on Agikuyu folk tales and proverbs.

Kiama retired to a life of farming in his highland home in Ngaini Village in Nyeri, Kenya. Two of his sons inherited his literary ability. One is a well-know journalist. Another son is Chief Editor of a major daily newspaper, known as The East African Standard.

VIGNETTE 6: NATURE AND LONGEVITY
The African Elephant

For those living far from nature, longevity is seen as an unnatural phenomenon arising out of modern technology. But those who know about the life of the African elephant, are aware that a fifty or seventy-year-old great-grandmother elephant is a vital resource for the survival of the species.

The oldest female in the group organizes elephant society through her daughters and granddaughters. The males are sent away at 14 years of age to live in their own bachelor herds, and are only allowed back for procreation. The male elephant keeps growing with a fifty-year old male achieving twice the size of a young mature female.

Male elephant society is governed a hierarchy based on brute strength, with some allowance given for the seasonal male heat cycle or "musth." Should two bull elephants in musth meet, then a most ferocious and deadly combat ensues.

In contrast to male society, female elephant society values old age, with its accompanying knowledge and experience. In times of catastrophic drought, which may occur at infrequent intervals, it is the grandmother or great-grandmother elephant who leads the family safely on long treks to distant lands in search of food and water. The matriarchal elephant can do this because she remembers their locations from her youth.

The caring of the young is a communal concern in which the older female relatives assist the novice mothers. By the same token, elder sisters practice breast feeding by letting younger siblings suck at their adolescent breasts.

By some means the female elephant family decides when their number becomes too large for their territorial ecology. A senior female member, her daughters and their young then leave the herd and go to some distant place and start their own new community. However, relatives still travel long distances to visit each other. When they meet, they show all the emotions of reunion. But the visitors eventually return to their own home territory after a few days of companionship.

Death in the elephant society—even of a new born baby— is mourned and a formal burial ceremony seems to take place. Even bits of dry elephant bones seem to be treated with what looks like reverence.

The elephant society I have just described is observed in a state of nature in which there is adequate land for the elephant. With increasing economic pressure, elephant land is shrinking and as is now well known the African elephant is an endangered species. Recent attempts to control fertility in the female by giving them the pill failed. The pill caused the females to give the wrong chemical signals to the males who would not leave them alone and interfered with the care of babies.

The elephant vignette illustrates that longevity can and does occur in nature. The vignettes on the highlanders' long lives show that with minimum technology Africans in some highland areas in Tanzania and Kenya can live for very long. However, the medical history of some of the children of these few men and women indicates that modern diseases such as diabetes and hypertension will not allow the progeny of these people to live so long. Furthermore, as is the case in the elephant society, rapid and unplanned urbanization and rural poverty are disrupting the central role of the elderly in the African family.

Under normal circumstances, in the central highlands of East Africa, longevity is seen as a blessing. "Old Man" or "Old Woman" is a term of great respect and status. There are even special greetings for the elders. The young long to achieve this status through the long process of growing old.

Thus, where the natural order and culture still prevail in Africa it is a great pleasure to grow old in the safety of your natural environment, surrounded by several generations of the African family, whether you are human or elephant.

COPING WITH OLD AGE IN AFRICA

Nana Apt

African Gerontological Association
University of Ghana
Legon
Ghana

INTRODUCTION

UBUNTU: I am because you are

UBUNTU is the communal African moral and political thought reflected in the communitarian features of the social structures of African societies. Jomo Kenyatta (1965) said: "According to Gikuyu ways of thinking, nobody is an isolated person....first and foremost he is several people's relative and several people's contemporary" The sense of community is reflected in many African proverbs: *The family is a crowd* and *One tree does not constitute a forest* (Ghanaian proverbs). In this paper, a clarion call is made to Africa to awaken to its own cultural values in developing models of care for the aging. Using past experiences, we will attempt to examine key issues of development, aging and quality of life and then make recommendations for effective social policies and programs that must be incorporated centrally into national development plans to create conditions of genuine well being for older Africans.

Presently, both developed and developing countries suffer from the strains and stresses of development. We should, however, be clear that we cannot solve Africa's problems by transferring models, institutions and techniques from other societies. There must be a harmonious relationship between the educational experiences of African societies and the major characteristics of the socio-cultural patterns in terms of beliefs, values, practices and needs, interests and problems.

Where has *UBUNTU* gone? An enquiry into the social situation and process of development must start with a determination of the position of the family in the face of modernization and the changes that are taking place. Traditionally, a complex interaction existed between the members of the family and that of a community. The economic, political and socio-cultural values of present-day Africa, with its sprawling urban centers and industrialized zones are increasingly different from its traditional values. The family is disintegrating and social change has brought a host of new social problems which remedial social welfare services are inadequately prepared to meet. In Africa, the concept of development has been greatly influenced by the colonial experience and Western modernization theory which posited that economic development and growth, mainly through

Longevity and Quality of Life, Edited by Butler and Jasmin.
Kluwer Academic / Plenum Publishers, New York, 2000.

industrialization, would automatically raise the living standards and meet the social needs of the population. In this respect, social welfare was viewed as a non-productive activity and therefore accorded a low priority in national development planning and resource allocation. In this context, the scope of social policy is limited and cannot deal with the critical problems of mass poverty and deprivation afflicting the majority of the people, especially in rural areas.

Until recently, the social arrangements of traditional Africa were, in the main, viewed simply as a barrier to economic development. The imperative was for modern social behavior and values as the necessary pathway to economic success. In this overly simple tale of what was required for the future of Africa, the positive economic functions of the African family system were forgotten. The argument presented here, and building upon earlier work (Apt and Grieco 1994; 1996) is that the African extended family system operated as a social welfare system. It has been hidden from history by intellectual and colonial stereotyping which viewed tradition in any form as a barrier to economic and social progress.

A consequence of this cultural blindness has been the failure to realize that the destruction of traditional relationships results in the destruction of the operational welfare system. Consequently, policies have been promoted both by national governments and international agencies. Their goal is radical change in existing social arrangements; however, no accompanying social welfare policies have been developed to compensate for the destruction of the traditional arrangements.

Here we argue, using the example of Ghana, that attention must now be paid to designing and implementing social welfare systems which are appropriate to the African reality. Economic development and urbanization have put the old social welfare system under pressure. At the same time there has been a general failure to develop alternative social welfare arrangements for those social categories which have lost out in this process of change. For example, advocating smaller families without policy thinking on elder care introduces other welfare problems. The destruction of social insurance relationships without the provision of a viable functional alternative generates both a social and economic crisis. The gap must be filled by new policy thinking before the traditional arrangements are put under such pressure that they collapse.

URBANIZATION: AFRICAN FAMILIES UNDER STRESS

What are the policies which contribute to the disintegration of the family and the traditional welfare system? In many African countries, government investment priorities have added to this disintegration, most particularly with respect to the imbalance between investment in rural and in urban areas. As in most countries of tropical Africa, the concentration of government investment in urban areas has widened the disparities of economic and social infrastructure between the country and cities, resulting in substantial rural-urban migration. Meanwhile, it is the rural areas which provide food and raw materials for the urban sector, despite increasing out-migration of the young.

In the wake of these demographic pressures, the family is disintegrating and social change has brought in its wake a host of new social problems which remedial social welfare services are not adequate to meet. Migration to the urban areas has contributed to high unemployment and overcrowding in the towns, as well as to the isolation of the elderly in rural areas where they become deprived of social and economic support. This trend if not seriously tackled, is expected to continue.

The problems of the elderly in the context of modern life are not confined to the rural areas. Increasingly, the indicators are of a trend away from the traditional perceptions and practices of obligations toward, and respect for the elderly. Support of the elderly in kind was a traditional practice of rural Africa; today, cash support of the elderly is increasingly a requirement of urban life. However, a small formal sector, low wages and employment insecurity work against urban Ghanaian offspring, who are unable to meet these income requirements of an urban life, as their rural counterparts are to meet the requirements of the rural elderly in kind (Korboe 1992). Clearly, the domestic separation of the urban elderly from the traditional family structure tells us something about the changing image of what the traditional family is and should be. Conflict of loyalties is evident between the newer urbanized conjugal family and the extended traditional family (Fortes 1971, 1981; Oppong 1981; Korboe 1992). In a study of the views of Ghanaian youth on aging (Apt 1991), it was evident that young families will not be living with their elderly much longer, as 81 percent of the youth interviewed were of the opinion that this arrangement was not feasible.

However, it would be a mistake to think that such separation is simply the outcome of the adoption of `modern' values and attitudes, for there are obvious infrastructural and structural factors involved in this change of practice. Urban housing conditions provide a major part of the explanation for these changes. In the rural Ghanaian context for example, the provision of accommodation for all social categories is largely not problematic; shortage of land is not a factor and simple additional dwellings are constructed of local materials as the need develops. In the urban context the provision of accommodation is radically more problematic. Low income accommodation is frequently found where there is little or no space for the erection of supplementary dwellings as families increase in size; urban accommodation typically requires cash payment to a landlord/lady and is frequently subject to the landowner's limitation on the number of persons entitled to inhabit a property. These factors taken together place pressure upon families to subdivide into component units (rural/urban), especially where family size is large. Such subdivision, in its turn, adversely affects the internal budgeting arrangements of the conjugal family with respect to its ability to meet its traditional welfare obligations.

It is not only the budgeting capabilities of the conjugal family which are affected by subdivision, for such subdivision has consequences for the arrangement of the various personal service and care arrangements within the extended family. For example, the traditional functions performed by the elderly with respect to child care are negatively affected by this domestic separation. Similarly, the caring services extended to the elderly with the traditional household become more problematic, sporadic and on occasion, even impossible, when elderly people become geographically separated from kin even within the same area of a city (Apt 1993). The reciprocity which existed between the generations in the traditional extended family is thus disrupted by urban life; in this process, the elderly who were previously valued for their services increasingly occupy the unenviable position of being viewed as useless consumers of scarce resources. Nevertheless, although the signs of an imminent crisis of social welfare with respect to the elderly in Africa are already visible, presently, due to the lack of a comprehensive social security system for all, the family continues to be the dominant source of care and support for the elderly in many African countries.

AGING AND QUALITY OF LIFE: WHAT ARE THE ISSUES IN AFRICA?

Longevity

Longevity is a cause for concern because of its consequence on the development of still embryonic social security systems. Few countries in Africa have public pension systems and where they do, these are typically restricted to a very small percentage of the older population, very often those who have enjoyed formal sector employment and have been public officials. The reality of limited resources in the face of growing family disintegration, and the growing trend toward nucleated families is becoming more and more apparent. On the other hand, there are no signs at the present moment of viable alternatives to provide care for older persons.

Low Income

Although there is a general consensus that the low income world that is normally representative in Africa does not have resources for catering to either the aging of society or large increases in the numbers of older persons through public sector provision #1 (World Bank 1994; Tout 1989), and although the policy focus has largely been on the family as the key agency in meeting the aging crisis (Phillips 1992), this simple opposition between the public sector and informal sector provision conceals a wealth of policy options which must be explored to make training relevant. The tendency to view the old simply as recipients of care and services is certainly no model for Africa and has served to conceal the prospects of, and opportunities, for mutual support.

The consequence of this current gap in knowledge about indigenous structures and values on aging with respect to Africa, leaves many key policy topics largely undiscussed and therefore untaught. Aging has not yet taken its necessary place on Africa's development policy radar. In discussions of older persons and development, older persons are simply regarded as constraints and not resources, and yet emerging information does not conform to this.

Social Change

In Africa, the older generations are living in a world which has undergone the fastest and greatest social and economic changes in history. Not only will there be increasingly larger numbers of older persons in our region but they have great educational needs which have to be studied and met if major human resources are not to be wasted. These educational needs range from literacy to technical training, through civic and political education, to banking and enterprise skills. Lack of education represents as great, if not a greater problem than physiological constraints to the social and economic participation of older persons in Africa. An increase in the number of generations in a fast-changing world necessarily accentuates the educational distance between the younger and older generations (Apt 1996) if education is viewed purely as the preserve of youth.

A lifetime approach to education which equips older generations with new skills narrows this gap. In regard to training, gerontologists have pointed out that older persons have considerable potential to learn. Therefore, training on aging in Africa must first and foremost begin with aging persons themselves.

Gender

Gender has received very little attention, although, after the Beijing meeting in 1995, there are signs that change is now on the way. The issue of gender and aging however has barely been addressed. Yet women progressively live longer than men in Africa and women more than men experience financial and cultural constraints which affect their quality of life as older persons. In Africa, women are the principal care and health providers within the family and women have fewer resources available to them to enable them to take care of themselves when they are old, divorced or widowed. Longevity in Africa clearly has consequences for female workloads and work opportunities if care for the oldest old is to take place within the family. It has consequences for the life opportunities of the care giver and stresses and strains are likely to take their toll on Africa's women care givers. The gender divisions in African societies which provide men with better access to employment than women, mean that for a woman the loss of a partner often constitutes a major change in her financial circumstances. Loss of a husband is often the loss of a way of life. There is a definite need for an aging policy in Africa to address the poverty problems of widowhood. Consideration needs to be given to creating income opportunities for older women in order to tackle their poverty. The aging of African societies necessitates a change in the traditional gender arrangements where women's financial position was mediated through their male partners.

Developing programs and policies which endow women with social and economic resources when they become widowed provides an opportunity for women to negotiate for better care and services within their own kinship structures. Thus, the longevity of women as compared with men raises the issue of developing appropriate forms of education and training based on gender. Gender and aging is clearly an area in which research and training activities should immediately be begun as part of a strategy to alleviate poverty. The domestic character of women's lives in Africa frequently leaves them unprepared for entering the public sphere on the death of a spouse. Thought must be given to enabling older women to take an active part in public life through appropriate training. Orientation and self esteem courses are necessary as routine features of women's education in their older years.

THE CHALLENGE OF EMPOWERING THE FAMILY

Africa's resource situation necessitates that she obtain assistance from outside agencies in delivering social welfare. While the contributions of the donor community are essential, it is important that such contributions be channeled into a cohesive welfare policy which is appropriate for us. The temptation to let the donors determine the direction of development must be resisted. But resistance is not sufficient. Donors must be advised about the limitations of many imported policies, to educate donors on the strengths of indigenous institutions and to develop with donors appropriate local solutions. There is an agenda which donors can follow in helping to develop appropriate social welfare policies for Africa.

- Donors have to encourage governments to directly target women, children and the aged in their social welfare designs.
- Donors must become aware of appropriate indigenous structures which can be harnessed in their programs rather than simply importing external technical models.
- Donors should seek to develop intergenerational project designs.
- Donors should focus on designing credit systems which are appropriate for women and for the vulnerable poor.
- Donors should ensure the participation of women in project design both as a method of effective project implementation and as a matter of good participatory practice.

CONCLUSION

The economic development of Africa cannot proceed further without more detailed and systematic policy thinking on social welfare needs, priorities and arrangements, not least because of the imminence of demographic changes. The comprehensive reconsideration and redesign of the classic social welfare systems of the advanced industrial countries, creates the opportunity to identify social welfare organizations which are appropriate for Africa.

That the family, even the African family, can provide all the social welfare needs of the society is surely a myth; that it provides the central structure of social welfare in a developing society is surely true. The time has arrived to design policies which support the family rather than destroy it.

Our suggestion is that policy be developed in such a way that it lends support to the traditional family-based social insurance arrangements; social welfare arrangements should not seek to replace the function of the family as they did historically in Western society but rather, endeavor to support and buttress the family in its performances of the important economic role of social insurance.

The family has been the cornerstone of social welfare arrangements in many African countries; contemporary policy-making must recognize this legacy and seek to incorporate it in the development of the national social welfare programs which are increasingly becoming necessary as Africa experiences a social welfare crisis. Affordable social welfare must make use of the traditional social welfare institutions. The policy question is: How can we best support these institutions and help them accomplish in the modern period what they accomplished so effectively in the past? The role of the state, given existing resource constraints,

must be one of support, not of substitution. The same relationships which were once viewed as barriers to development must now be seen as an essential source of support for the development process. The main thrust in the argument hinges on inadequate and inappropriate research on the family. Although research designed to fill gaps in knowledge is very urgent and important, high priority should be given to research oriented to the specific problems of the African region including methodological studies. Such research is best carried out in African countries themselves and by competent persons with knowledge of national and regional conditions. The following areas require research in order to fill existing gaps in knowledge:
- The social, cultural and economic determinants of population variables in different developmental and political situations, particularly at the family and micro levels;
- the demographic and social processes occurring within the family cycle through time and, particularly, in relation to alternative modes of development;
- the interrelations of population trends and conditions and other social and economic variables, in particular, the availability of human resources, food and natural resources, the quality of the environment, the need for health promotion, education, employment, welfare, housing and other social services and amenities;
- (2) the enhancement of the status of women, the rights of children and the elderly and (3) the need for social security and political stability;
- the impact of the change in the size of families on biological and demographic characteristics of the population;
- the changing structure, functions and dynamics of the family as an institution, including changing roles of men and women, attitudes toward opportunities for women's education and employment; the implications of current and future population trends for the status of women; biomedical research on male and female fertility, and the economic, social and demographic benefits to be derived from the integration of women in the development process;
- development of social indicators, reflecting the quality of life as well as the interrelations between socioeconomic and demographic phenomena should be encouraged. Emphasis should also be given to the development of socioeconomic and demographic models.

If society expects the family to serve as the primary safety net it once was, it is necessary that more be known about how social changes have affected the family's ability to undertake such responsibilities. Developing an empowerment perspective, a movement away from the Western medical paradigm of deterioration, enables older persons within the family confines to define an active role for themselves in social and economic life. Similarly, mutual support structures, so effective in traditional African societies, provide older persons with more control over their own lives.

Although there is a general consensus that the low income world does not have resources to cater to either the aging of society or large increases in the numbers of older persons through public sector provision (World Bank 1994; Tout 1989), and although the policy focus has largely been on the family as the key agency in meeting the aging crisis (Phillips 1992), this simple opposition between the public sector and informal sector provision conceals a wealth of policy options which must be explored. The tendency to view the old simply as recipients of care and services is certainly no model for Africa and has served to conceal the prospects and opportunities for mutual support. The *UBUNTU* philosophy which pervades Africa's indigenous culture must be revisited. The consequence of this current gap in knowledge about indigenous structures and values on aging with respect to Africa, leaves many key policy topics largely undiscussed and therefore untaught. Aging has not yet taken its necessary place on Africa's development policy radar. In discussions of older persons and development, older persons are considered constraints, not resources. Yet emerging information presents a contrary view. The need to empower older people is obvious. But the empowerment of older people is not only about rights: it is also about obligations. Conditions must be created to enable older persons to exercise maximum autonomy, and to put fewer demands on scarce social resources. This area needs extensive study.

Lack of education represents as large, if not a larger problem than physical constraints to the social and economic participation of older persons in Africa. An increase in the number of generations in a fast changing world necessarily accentuates the educational gap between the younger and older generations (Apt 1996) if education is viewed purely as the preserve of youth.

Finally, let me end as I started by reaffirming my belief in *UBUNTU*. For the large majority of African countries, mutual support schemes should be the more sustainable alternative to a blanket social security provision. Tout (1989) discusses precisely these issues in his proposed new approach to the problems of older people in developing countries. Mutual support however should not be viewed simply as the cheaper option but should be viewed in terms of the active participation benefit it brings. Mutual support offers the opportunity for enhanced sociability, greater bargaining power across the generations and greater security in the context of shrinking kinship structures.

Older persons, with their greater endowments of available time, can make major contributions to education and training. The notion that education is for the young in Africa must be combated. Assuming that classrooms are only for children closes the door of opportunity for older persons to gain literacy. Besides, the generational skills of older persons, such as traditional craft and cultural history, can be brought to the classroom and enhance the education process. Thus, designing an appropriate role for older people in the curricula of Africa's young generation is a policy step worth investigating.

In conclusion, an examination of the opportunities and challenges of aging and longevity in Africa, what it means and what desires it connotes, cannot neglect the role, aspirations and active involvement of old people in their environment. Grassroots action is required to educate and inform both older and younger Africans caught in transition, that they are capable of taking individual appropriate indigenous actions and arriving at new social roles.

REFERENCES

1. Apt, N., 1996 Coping with old age in Africa. Avebury, Aldershot.

2. Apt, N. , 1991 Ghanaian Youth on Aging, BOLD Vol.1: 3.

3. Apt, N., Grieco, M. 1994 Urbanization, caring for elderly people and the changing African family. The challenge to social policy. International Social Security Review Vol. 47, 3-4.

4. Ayisi E. 1972 An introduction to the study of African culture, Heinerman, London.

5. Ellis, J. Ed. 1978 West African families in Britain, Routledge, London.

6. Fortes, M. 1971 The family: Bane or Blessing, Ghana University Press, Accra.

7. Fortes, M. 1981 Foreword in: C. Oppong, Middle Class African Marriage: A study of Ghanaian Senior Civil Servants, pp ix-xii, Allen and Unwin, London.

8. Grieco, M.& APT, N. 1996 Interdependence and Independence: Averting the Poverty of Older Persons in an Aging World. Bulletin on Aging 2 & 3, United Nations, New York.

9. Kenyatta, J.1965 Facing Mount Kenya, Vintage Books, New York, p 297.

10. Korboe, D. 1992 Family houses in Ghanaian cities: To be or not to be, Urban Studies Vol. 29:7.

11. Oppong, C. 1981 Middle class African marriage: A study of Ghanaian civil servants. Allen and Unwin, London.

12. Tout, K. 1989 Aging in developing countries. Oxford University Press, Oxford.

13. World Bank 1994 Averting the old age crisis. The World Bank, Washington.

3. Women and Longevity

WOMEN IN THE LONGEVITY REVOLUTION

Betty Friedan

2022 Columbia Road
Washington, D.C. 20009
USA

Women's power is basic to the longevity revolution in more ways than the obvious fact that the great majority of people over 60 in the next few decades will be women. Women are outliving men by nearly eight years both in the United States and in most post-industrial societies. And, as women stop dying in childbirth and get control of their own reproductive process, this will also occur in the developing countries. Nobody is paying much attention to this. If I were a man, the question I would ask is why millions of dollars aren't being spent on research to find out why women are living longer than men. Whenever I have lectured on this to a mixed audience, it makes men very uneasy. I don't know why. Maybe they don't want to admit that women are living longer, that there could be a real superiority there. But we might say: If only men could be more like women, they might live longer!

In any event, women will be the power players in the longevity revolution, but it is a new kind of power. This power was evidenced in the women's movement that has transformed society in this last half century in America and in France. It is now spreading throughout the world. There is no question of that, even though in Muslim countries, for the moment, women are being pushed back under the veil. But the world is such a small village now that I don't think it will work anymore, anywhere, to take back from women the freedom, equality of opportunity and the control over their own destiny that they now take for granted.

Why is the kind of power that we see in the emergence of women getting control over their own lives and destiny basic to the longevity revolution? I can remember many years ago, at the first population conference in Bucharest, when the women's movement in America was exploding. I was brought in to give advice to John D. Rockefeller who was leading that population conference. I said to him and to the other population people: No matter how many IUDs or other forms of birth control you spread around the world, nothing will get the population explosion under control unless women are given other forms of security for their lives than

Longevity and Quality of Life, Edited by Butler and Jasmin.
Kluwer Academic / Plenum Publishers, New York, 2000.

sons. Women must receive education and equal employment opportunity and *then* the right to control their own reproductive process, to decide why, when and how many times to bear a child.

The second aspect of this new kind of power is life-affirming. I came across statistics from the Midtown Manhattan Study and the National Center of Health statistics that related mental health to many variables. This study, some thirty years ago, found that women's mental health in America peaked at 20, decreased every decade after 20, and drastically after 40, compared to men. As we now know there is an enormous mind-body connection. The kind of depression women were experiencing was so endemic at that time in America that it was actually called *involutional melancholia.* It was considered virtually normal for women to be hospitalized for depression when they stopped being nubile; when their reproductive cycle was over. About twenty years later, after the women's movement, the studies were repeated. To everybody's amazement, *involutional melancholia* was no longer occurring. Women's mental health was as good or better in their 30s, 40s, 50s or even 60s than it was in their 20s. In fact, there were new signs of pressure on young women in their 20s. They had a lot of decisions to make. They were beginning to enter careers and marriage was not the only way to go. But this kind of stress can't be compared to the fact that they were no longer depressed at 40 or 45. This enormous improvement after 40 was not happening with men. I do not think that men want to admit that the great majority of survivors after 60 will be women unless something different begins to happen among men, and that is where the frontier is.

What has to change? Others at this conference have already mentioned that major basic research done by Bob Butler, James Birren and others in the study of aging show that the two most important factors to vital longevity are not diet, exercise, vitamins or hormones. Rather, they are a sense of purpose and social bonds. Purpose in life that structures a person's days and gives one a reason to live, and bonds of intimacy with others, with friends and family. Love and work.

Questions have to be asked–and some gerontologists have begun to ask then. For example, if love and work are the most important things, do you have to love the way you loved when you were 20 or 30 or not love at all? Do you have to work the way you worked when you were 20 or 30 or not work at all? Do you need the power that you thought you needed when you were 40 or do you just slump into a powerless passive victim state? Those questions are just beginning to be asked.

It has been said here, and I think it is true: In this new era of the longevity revolution, we will have to see new kinds of heroines and heroes and new kinds of love stories that are not based on youth or the imitations of youth. Viagra evidently will make men studs until they are 100. This may be a fad that dies out like the hoola-hoop! It may just be a temporary obsession. I think it is no more cutting edge than the fact that we now hear that women can give birth at 50-60-65. Women may have the option of giving birth at 60, but I doubt that many will choose to do so. We have to move toward a sense of personal empowerment, evolution and growth beyond the values of youth.

The longevity revolution does not mean that older people will now become young. It's a revolution in which we will see an evolution of human life. We are beginning to see some hints of this. Research shows, for instance, that the secret of a long vital age comes from people who have combined many roles in life, not just the role of housewife or single career. When you look at women's experience, part of their flexibility in old age and their ability to respond to change more easily than men, results from the discontinuities in their lives which give women enormous strength. And men will have to acquire that strength. Nobody will be able to live through this long longevity revolution with a single career or a single stage of family life. Women are already moving beyond the reproductive role that comes with their biology. They now have a third of life left to live after that role is finished. Men are not going to have careers that are in a single line, with one company or only one profession. There will have to be a continual reinvention of ourselves, our purposes, our careers, the mixes of career-work-children-family that differ as life goes on. Other forms of values will become more important as well, in terms of personal growth, creativity, service to the community.

In my country there hasn't been a move for a shorter work week for fifty years. I'm going to try to start one. My next revolution will be against a culture of greed and excessive worship of the almighty bottom line, the dollar, the GDP. In the longevity revolution, women and men will have to liberate themselves from the culture of greed. The Quality of Life (QOL) must become more important than the GDP. We are in the middle of a paradigm shift as we approach this longevity revolution.

We are seeing women who lived through the era of what I call the "feminine mystique" They were supposed to live their whole lives as housewives and mothers and then it turned out that they were living so long that they had to ask themselves, "What am I going to do when my children grow up?" Women returned in great numbers to community colleges and universities, where the average age for undergraduates, in the United States at least, is no longer 18-19-20. It is 30 and older. People in their 60s are going back to school, to college for the first time, and women who would have been called "matrons" in another time are starting a new life in community colleges. They are revitalizing classrooms in a way that's reminiscent of the veterans who came home after World War II. The role of the student in midlife or in later life will become increasingly common.

The nature of women's empowerment and what it has done to transform society is pertinent to the longevity revolution. The Eagleton Institute at Rutgers, New Jersey, studied women in politics. The election of as many as two women to a state legislature began to change the agenda of that legislature, away from the rule of the almighty dollar, to measures that embraced the needs of life, of children, of old people, of the sick–not just women's rights but quality of life issues. As we move into the longevity revolution in all our societies, the participation of women will change the agenda. In my country, we have studied something that we call the gender gap and it has never been larger. Women in America no longer vote the way their husbands do. The women's movement gave them a consciousness of their own power, and of their desire to make the decisions that control their own destiny.

At this moment in time, as we become more acutely conscious of the longevity revolution, we have to realize that it is a different kind of revolution. Hannah Arendt in her famous book on revolution said that "most revolutions are done by a group that has been oppressed, taking the power away from the oppressors and then oppressing the oppressors in turn." She said the American revolution was different because it emerged from a democratic process that was then built into the society. In the same way, the longevity revolution is not going to evolve into one group taking power away from another group. Rather, it is going to transform the nature of power.

One might predict that in the future women will rule because they are going to be living longer. I don't think that is what will happen. Women freed themselves from stereotypes that prevented them from using their full potential in society. Consequently, in the future we will no longer have such blatant sexual stereotypes of either gender. But the process will not be complete until the child rearing that has been seen as women's role, is seen equally as men's role. This is beginning to happen. A few years ago *The New York Times* published a front page story: American men were not carrying 50 percent of child care and housework. I thought it was splendid that *The New York Times* could even conceive it possible that American men *would* share 50 percent of housework. In fact according to the latest statistics, men are now carrying nearly 40 percent. They are not doing the ironing, but neither are women so the whole thing is changing.

Maybe men's growing participation in child care will give them a sensitivity to life and an appreciation of its importance. Perhaps it will enable us to redefine the prevailing values in our societies in terms of quality of life, rather than being dominated by the culture of greed.

Research on age done twenty years ago seems to show that in mid-life there is a crossover: men become more like women, women become more like men. Today, when women have the courage to be assertive earlier in life and men are sharing the child rearing, which requires tenderness, maybe we will not have to wait until mid-life for the crossover to occur. As life evolves, what has been considered masculine and feminine is now somehow united in the totality of human evolution, selfhood and experience. As we move further into the longevity revolution, I predict the importance of gender differences will diminish still further. We will move on to a new stage of human evolution. Human values that influenced the so-called feminine sensitivity to life will be shared by both men and women.

I believe that as women take on more of their share of the power in all our nations, and as men are sensitized by the reality of caring for their young, the idea of nuclear weapons will become impossible. There will be no longevity revolution if we endanger our earth with nuclear weapons. I see two poles: one is a dangerous proliferation of nuclear weapons and the other is the longevity revolution. I hope the longevity revolution prevails.

Political Responsibilities in the Longevity Revolution

HOW SEVEN EUROPEAN COUNTRIES HAVE ANTICIPATED THE CONSEQUENCE OF POPULATION AGING IN THE HEALTHCARE SECTOR: A SURVEY OF ERNST & YOUNG

Pierre Anhoury

Ernst & Young Sante
6 place de L'iris
Paris La Defense II 92095
France

INTRODUCTION

Within the next two years, four European countries (Italy, Germany, UK and France) will each have more than ten million people aged 65 and older. In addition, the over-75 age group is the fastest growing population. The vast majority of people over age 65 are low-income women. The needs of this population are significant. Life expectancy of people without disabilities is 73 years for men and 74 years for women. This fact should be taken into account by healthcare programs, which should not only focus on the issues of mental illness and accommodation.

METHODOLOGY

Ernst & Young conducted a survey in seven European countries. We selected the countries because of their respective cultures and the initiatives that have been taken in this field. The survey was coordinated in France. Our team began by performing a preliminary analysis of the available data in order to provide each member of the

Longevity and Quality of Life, Edited by Butler and Jasmin.
Kluwer Academic / Plenum Publishers, New York, 2000.

team with the same information. Then, the team interviewed five experts in each country (demographer, economist, member of a government, geriatrician, member of an association). We asked questions in the following three areas:

- How did health care organizations in your country anticipate the impact of aging? What unique and innovative actions have been undertaken?
- What initiatives did the organizations take in order to face the foreseeable increase in costs?
- What was the reaction of the pharmaceutical industry to the proposal that products or new services be adapted to these age groups?

Ernst & Young performed a detailed analysis of the answers to these questions. During a seminar, each of the seven countries received a score ranging from 1 to 10, according to the country's preparedness for the consequences of aging. In addition, the team gave recommendations for the main themes that emerged.

The results of the survey are presented in the order the questions were asked followed by the recommendations given. The answers do not refer to a specific nation, unless one specifically needed to be mentioned.

RESULTS

Nomination of a person within the healthcare department who will be in charge of aging

Governments usually rely on research organizations, universities and demographic institutes when it comes to establishing missions. Within the governments, no department is in charge of analyzing the impact of aging and the measures that need to be taken.

As the stakes are high, we believe that it will be necessary to nominate a well-known coordinator. This person will be in charge of centralizing the national and European information, will legitimize the initiatives taken and honor aging, which is too often seen as an unfortunate condition that needs a simple palliative treatment. A pro-active national policy could then be put into place.

Evolution of healthcare structures and the methods of care

Healthcare establishments in all countries have experienced or anticipate a decrease in the number of beds for short-term care to the benefit of average and long-term care. This evolution is not completed and proceeds with variable speed according to the country. The financing arrangements sometimes block this evolution. For example, in France, an elderly person who is accommodated in a bed for short-term care will have the fees reimbursed, whereas, the same stay in an average or long-term care bed involves significant fees for patient and family. The quality of the physical surroundings and the services of these establishments should be improved.

We think that it is necessary to introduce financial incentives and in particular heavy tariff incentives so that restructuring accelerates. These incentives will also allow the institutions to offer a total number of beds that are better adapted to the needs of elderly people.

Training of healthcare professionals

The training of healthcare professionals (nurses, doctors) focused on the care of the aged varies considerably by country. All persons interviewed called attention to the need for specific training. However, nowadays, the content, length and qualifying character of geriatric training differs. Geriatrics is not recognized as a specialty and is often seen as secondary training.

We think that it is necessary to build a European program that will address doctors and one which delivers a diploma that is acknowledged by the countries of the Union. This same initiative needs to be taken for nurses for the same reason as applies to specialty training for anesthesiology or surgery.

National or regional measures of accompaniment

The census of measures taken in each country shows diversity of information which merits being shared. All countries have demographic studies at their disposal. In addition, they possess studies on the state of health of elderly populations. The dominating themes differ (dependence, psychiatric troubles, cardiology problems, cancers, nutrition, etc.).

Some countries, such as France, have noted the improvement of buildings. Other countries, such as The Netherlands, have brought up the methods of care at home. Finally, a country such as Denmark has balanced the effort by taking an interest in healthy old people (office of aging and sports). Associations have been created in two countries. They are dedicated to health problems like incontinence or Alzheimer's disease. Finally, initiatives have been taken that aim at financial support for aged dependents. The methods of the amounts deducted vary (taxes, contributions, etc.).

It is very difficult to compare financial efforts on behalf of elderly people. Some classify this effort in the health services category. Others categorize it inside the family or social category. These three categories differ from country to country and it would be incorrect to make comparisons.

We think that a European agency could collect this information and facilitate the dissemination. In the same way, consistency in the rules of distribution of means (health, family, social) would finally allow each country to measure its effort in comparison to others.

Innovations in the field of prevention

The prevention of the consequences of aging was the object of interesting national programs. The UK distinguishes itself by having developed programs to fight against amnesia. Denmark developed actions in the fields of nutrition, sports and information relating to the elderly. Germany put its effort in the field of post hospitalization for short term care and the importance of the role of volunteers. Italy has particularly developed the exchanges of information between hospitals, after-care structures and doctors (town), and non-profit initiatives.

We think it is necessary that nations review these initiatives so they can share information and learn from each other.

Measures taken to control current and future expenses

All countries interviewed have taken initiatives to control the escalation of costs entailed in maintaining the aging population. In addition to the growth of receipts, the analyzed countries have limited the reimbursement of services that they consider to be less important and favor the use of generic medications.

In Denmark, preventive medicine is expanding and Italy and France insist that the general practitioners be the first point of contact for an older person.

The establishment of networks that rely on structured links between doctors, care establishments and patients (the "Socrates" model in Germany) is eagerly anticipated. In addition, we seek the best care continuum between ambulatory medicine, the stay in hospital and aftercare.

Development of the market of medical equipment dedicated to elderly people

The medical equipment market dedicated to elderly people is expanding and is most evident in Switzerland. This equipment includes articulatory prostheses, wheelchairs, equipment to combat articulatory deformations, hearing aids, dental prostheses, etc. The medical equipment market involves an industry of small and medium-size companies, which are spread all over Europe and subject to European Community regulations. Comparative valuations of quality, security and price need to be established.

We recommend that a study be conducted at a European level by an independent agency, such as the independent American institute ECRI, which specializes in the valuation of medical technologies. In 1990, all wheelchairs, canes and prostheses were deposited at the institute. These comparative tests gave rise to publications which are used as reference. In addition, they have led several manufacturers to make considerable improvements.

Reactions of the pharmaceutical industry facing the aging of populations

The aging of populations is the growth engine of the pharmaceutical industry. No matter what price and volume control measures are taken, more and more people will be required to take medicine.

Presently, the industry's approach remains traditional in the seven countries analyzed: Offer new products and prove their advantage, sell them at the highest price and favor the highest possible consumption, no matter what the age of the consumer.

Today, payers are questioning the pharmaceutical industry and trying to collaborate by getting agreement on more precise indications for medications, on actions which focus on getting a better follow-up of treatments, and stricter adherence to necessary limitations. These innovative approaches can be found in particular in Germany and Italy. Hopefully, they will spread rapidly.

We encourage the creation of an authority to bring together payers, suppliers and consumers for the coordination and sharing of experiences, and to allow for the introduction of therapeutic innovations in each country.

CONCLUSION

The analyzed countries reveal numerous helpful and often unknown initiatives, but an awareness of the consequences of aging varies considerably between countries.

Some countries are still exploring options, while others have working programs in place. These latter countries received scores of 7 out of a possible 10. Our group rates the countries as follows:

Denmark and The Netherlands have each taken the greatest initiative and are also the most structured. They each received a score of 7. The United Kingdom and Germany each received a score of 5 because they are proceeding at a slower pace. Switzerland, Italy and France each received a score of 3 out of 10 because they anticipated the least and their responses remain in the experimental phase.

The situation needs to change rapidly. The four countries who will have the greatest population of persons over 65 in the year 2000 are those who received scores of 5 and lower. Programs must be set in place quickly to meet this challenge.

RESPONSIBILITIES OF THE INDIVIDUAL AND SOCIETY: POLITICAL RESPONSIBILITIES IN THE REVOLUTION IN LONGEVITY

Ursula Lehr

University of Heidelberg
Ruprecht-Karis
Institute for Gerontology
Bergheimer Str. 20
Heidelberg 69115
Germany

INTRODUCTION

A policy for the aged should be based on a valid diagnosis and a careful analysis of the ways to cope with the challenge of longevity. A valid diagnosis of the consequences of longevity must discuss the phenomenon in the context of demographic change.

THE DEMOGRAPHIC CHANGE

The demographic change caused by the rise of individual life expectancy on the one hand and the decrease in the birth rates on the other, is a challenge to society and the economy, to politics and science, as well as to families and individuals.

The life expectancy for a newborn child today in Germany is 73 years for males and 79.9 years for females. But, a 60-year old person can expect to live 21-22 more years. That means, after retirement a person will live about 20 more years, one fourth of his/her life.

We are living in a graying world. One hundred years ago, the percentage of persons 60 years and older in Germany was 5 percent, today it is 21 percent—18 percent of all men and 24 percent of all women. In the year 2030, 37 percent of the

Longevity and Quality of Life, Edited by Butler and Jasmin.
Kluwer Academic / Plenum Publishers, New York, 2000.

population in Germany will be 60 years and older, and there is also an increase in the 70, 80, 90 and 100-year-old age groups. Twenty years ago we had only 385 centenarians in our country; in 1994 our *Bundespräsiden*t congratulated 4,602 persons (582 men and 4020 women). In the year 2000 we will have more than 10,000 centenarians in our country. Scientists of all disciplines and faculties, administrators, practitioners and politicians must discuss the question of longevity, and that of psychological and physical well-being. What can be done to assure healthy aging? What can be done to assure the quality of life in old age?

The proportion between the different age groups has changed. One hundred years ago, the proportion of persons younger than 75 to those above the age of 75 was 1:79; in 1925 the ratio was 1:67; in 1950 it was 1:35; in 1970 1:25; in 1994 1:14. It is projected that in the year 2020 it will be 1:8 and in the year 2040 the proportion will be 1:6.

We have seen an enormous change in the structure of households. There is a significant trend away from the three or even two-generation households, toward the one-generation household. Only one percent of all the 34 million households in Germany are three-generation households. Of all people 65 and older, nearly 40 percent are living in a one-person household; of all women 75 years and older, 68 percent; of all men 75 and older, 28 percent. Four percent of the elderly aged 86 are living in one and two-person households. The combined effects of smaller family size and increased female labor force participation, and the increasing longevity (sometimes with two retirement age generations) means that the family's ability to serve as an important social institution for the care of the elderly can quickly erode.

We also have a trend away from the three-generation family to the four-generation family, all alive at the same time, but not living together. Today a newborn child very frequently comes to know all of her/his four grandparents, and very often two great-grandparents, too. Persons sixty years and over with great-grandchildren are quite common (about 20 percent), as are persons 60 years and older who are caring for their own parents (also about 20 percent). It means that caring for an elderly family member very often is done by grandparents.

It also means that family care has its limitations. Today 75 percent of the disabled elderly will be cared for by family members, mainly wives or daughters; about 10 percent will receive care from family members, together with professional caregivers, and 15 percent exclusively from professional caregivers. Four percent of the elderly are living in institutions. But this will change in the near future, because of the care insurance we have introduced in 1995 which pays family members to provide care in private households. It has motivated many family members to assume this difficult responsibility.

However, the longevity revolution should not be discussed primarily in the context of care and caring. A representative study of INFRA-Test (1992) on the prevalence of dependency in old age, involving more than 26,000 households, shows that aging in no way automatically results in dependency. In the age group of 60-74 years old, only 3-4 percent were dependent and needed help. In the age group 75-85 years, only 10 percent were dependent, indicating that 90 percent of this group were fully capable of meeting the challenges of daily life. In the group 85 years and older 21 percent of all men and 28 percent of all women were disabled and dependent. In other words, 79 percent of the men and 72 percent of the women 85 years and older could manage their daily lives.

On the one hand, the need for long-term care is increasing, primarily determined by demographic developments. The trend toward single-person households and small families will continue and the number of people over the age of 80 will rise. Thus, the number of persons in need of long-term care may well increase. On the other hand, according to research findings of Svanborg from Göteborg University in Sweden, of Manton from Duke University in the United States and of the Gerontological Center of Geneva, Switzerland, older persons today are in much better health than the aged were 10 and 20 years ago. The estimated number of persons needing long-term care in the future could be lower if the increase in life expectancy is the result of a healthier lifestyle, prevention and rehabilitation.

There is a trend away from a three-generation contract to a five-generation contract. The idea of a three-generation contract is that those persons in the labor force must provide (via insurance contributions and taxes) for those who are not yet working and for those who have finished working and are retired. Around the turn of the century, the 15-70 year olds provided for those 15 and over 70. But at that time just two percent of the population was 70 years and older. Today, people aged 25 to 60 in the labor force provide for those who are not yet working (in some cases this includes two generations because some 30-year-old students have their own children in kindergarten) and for those who have already stopped working. These represent 26 percent of the population, often two generations, with consequences for our pension system. Today the economic situation of the elderly in our country is still good; only 1.5 percent of persons 60 years and older receive social benefits (mostly women, 75 years and older). From the 3.9 percent of the elderly living in homes for the aged, nearly 40 percent need social welfare support. In our country it is quite unusual that grown children give important financial and material support to their parents. Very often elderly parents and grandparents support their children and grandchildren from their pensions.

A policy for the elderly should be more than a care policy, and it must include more than financial considerations. In order to cope with the challenges of longevity three issues should be stressed in a policy for the aged:

- Ways to maintain and increase the competence of the elderly in order to prevent dependency and to secure a healthy aging
- The extension and improvement of rehabilitation.programs for the aged that enable them to continue to live independently
- Problems of the dependent and frail elderly, and the issues of care

A POLICY FOR HEALTHY AGING

A policy for the aged is a policy for healthy aging, combined with a state of psychological and physical well-being. Aging is affected by biological heredity and individual behavior, as well as a wide range of social, environmental, cultural and political factors. Healthy aging is the result of a lifelong process. We know that a variety of influences determine the processes of aging and of well-being in old age. They include early childhood, adolescence, early and middle adulthood, as well as the present life situation of the aged. Interdisciplinary cooperation is needed, to which the biology of aging, the field of geriatrics, and the behavioral as well as social sciences contribute. Healthy aging is a challenge: for gerontologists, scientists of many disciplines, and politicians, too.

Preventive Measures to Maintain and Increase the Competence of the Aged

What can be done to secure a high quality of life in old age? Obviously, hygiene, preventive care and an overall good health care system are needed. But it is also important to inculcate healthy habits in the young.

There are many studies which confirm that physical activity is a prerequisite for successful aging. Age-determined physical changes, such as functional impairment of organs and changes in the motor, muscular and respiratory systems, are similar to the effects of lack of exercise in the young. The young, physically inactive individual seems old, just as the old but active individual seems young. There is a slogan: "Sports help you remain 40 for 20 years." Physical activity also affects certain mental abilities, subjective well-being, social skills and self-confidence.

Mental activity is a prerequisite for successful aging, too. Many studies, especially longitudinal studies, have found that mentally active people, that is, individuals with a relatively high IQ, a wide range of interests, a far-reaching future perspective and social contacts, reach old age with greater feelings of emotional and physical well-being than individuals who lack at least some of these attributes. It has been established that cognitive activity is essential for healthy aging. An older person must remain mentally active; he or she must be furnished with (new) information and be intellectually challenged. Reduction in mental activity can speed up the process of aging. This can also work in reverse: intellectual challenges can result in older people acting in ways one would only expect from younger persons.

Motivation for mental, physical and social activity must be increased. Numerous possibilities must be offered in order to make this goal attractive to older individuals. Many barriers which are built into our society (society's negative image of older people, conditions such as limitations of access to transportation, etc.) must be reduced or eliminated.

The importance of physical and mental activity must be emphasized from childhood. Men and women must also remain active as adults. In this context, problems of lifelong learning and opportunities for occupational activities for women should be addressed. Occupational activities after retirement are important preventive measures that help maintain health and competence in old age, as well as contributing to society. Retirement options must be kept flexible from person to person, permitting earlier as well as later retirement. This should be discussed from medical, psychological, sociological and ergonomic points of view. In the future, it is doubtful if we can afford to have one-third of the population retired. For many, it can mean an accelerated loss of abilities and competencies. *Fördern durch Fordern*—"to promote by challenging"—should be our motto.

Another kind of prevention is the development of coping strategies. A person needs to learn during the whole of life how to cope with stress situations.

Measures of Rehabilitation

The process of healthy aging can often be facilitated by rehabilitation measures and the management of problem situations. The decline of abilities can frequently be avoided if activities and competencies are initiated immediately after a crisis situation arises.

Psychological research contributes to the foundation and practice of geriatric rehabilitation by different approaches and findings which can be summarized.

The falsification of the deficit-model

The last thirty years of psycho-gerontological research has provided evidence for the secondary role of the age factor in determining constancy or change in cognitive and social functioning. Socioeconomic factors, personality, health and environmental contributions to the functional status of the aged contribute to a greater degree than chronological age. Experimental research on the effects of intervention strategies have demonstrated a high degree of plasticity. From this point of view rehabilitation measures for the aged are urgently needed.

Conclusion: Effects of disease, deficits in cognitive and psychological functions, which formerly were regarded as irreversible can be reduced by adequate intervention.

Distancing oneself from wear-and-tear theories

Activity is one of the most important conditions for the preservation and restoration of cognitive, psychomotor and social functioning in old age. Wear-and-tear theories suggesting rest and inactivity for prevention and treatment are disproved by many findings.

Conclusion: Prevention and rehabilitation should therefore focus on the activation of physical, mental and social abilities.

Patterns of aging versus age-norms

Research on aging has provided evidence for the existence of a variety of patterns of aging. These patterns may have some biological sources, but more important are personality and social, economic and cohort-specific conditions. The multi-dimensionality and multi-directionality of aging processes have to be taken into account.

Conclusion: In addition to the medical point of view, rehabilitation, psychological and social aspects should be taken into consideration.

The individuality of processes

Aging is a highly individual and highly specific process, not only for each individual, but also for different organ systems within the same individual. From a psychosocial point of view the biography of an individual must be taken into account when planning a rehabilitation program. In this context, historical events are important, especially the age at which a person faced these events (war, inflation, currency reform, displacement, etc.). Many attitudes and behaviors which are age-related (such as religious beliefs, attitude toward sex or political behavior) actually are cohort-specific variables and may change in future cohorts.

Conclusion: Measures of rehabilitation should be individually designed and take into account the patient's biography.

The role of social networks

The decisive role of the social network in coping with physical handicaps and mental problems was demonstrated in many studies. "Significant others" can be helpful in some situations. But they also can contribute to a reinforcement of dependency.

Conclusion: The family, partners and other persons in the social network should be integrated into the rehabilitation process. These persons should be given information and, if home care is needed, qualifications required for these tasks. Members of the family should be given care, too, in order to prevent a breakdown of their own health.

The role of ecology

The role of ecology in the shaping of human behavior and experiences was already demonstrated 20 years ago. Lawton (1970) formulated the "docility hypothesis," according to which the variance in independent lifestyle explained by ecological factors decreases with increasing competence of the person. This means, however, that with increasingly poor health ecological factors gain increasingly more influence. Very often ecological variables as defined by the housing situation, the degree to which the home is adequately equipped, rural or urban environments, transportation, etc., will determine the degree of dependency of a person. Favorable ecological conditions extend the range of the activity of the person and elicit more activity, social contacts and provide more stimulation. Unfavorable ecological variables may result in unnecessary restrictions and a loss of abilities and skills.

Conclusion: Rehabilitation plans must take into consideration ecological factors. Often an adaptation of the home to the specific needs of an individual will facilitate more independence for a handicapped elderly person. General guidelines for the construction of apartments for the elderly will be of limited help in this context.

The role of cognitive representation

Many studies have shown how cognitive representation shapes human behavior. The individual behaves according to the situation as perceived rather than according to the objective situation. This is true for a person's behavior toward family, the housing situation, and health. Physicians should take this into account in their interactions with older patients. It is most important for the subjective health of the patient. In the Bonn Longitudinal Study on Aging it was shown that subjective health was more closely associated with longevity than objective health. Subjective health was also a signficant predictor of mortality in the New Haven Health Survey.

All members of the rehabilitation team should be informed about the role of the patient's subjective awareness of the situation, especially health status. Too much or too less information can have a detrimental effect on the well-being of the patient —depending of course on the person and the specific situation. These effects can retard the rehabilitation process.

Conclusion: Doctors and health care teams should know that behavior is dependent on the perceived situation rather than on objective reality.

Patterns of coping with health problems

The role of patterns of coping with poor health was demonstrated in studies on patients suffering from stroke, cardiovascular diseases, cancer, etc. The success of the rehabilitation process was shown to be dependent to an equal degree on the patient's ability to effectively cope both emotionally and physically with illness on the one hand, and on the improvement of functional (ADL) status on the other. Favorable coping strategies included acceptance of the disease or the handicap (but "making the best of it"), which was followed by achievement-related behavior, using opportunities, optimism and "adjustment to the situation." Unfavorable response patterns included evasive reactions, depressive reactions, noncompliance and "aggressive behavior." These unfavorable coping styles were found more frequently in patients who believed in the unchangeability of their situation and in a poor sense of control of this situation.

Conclusion: The patient should be informed about ways to deal with his/her situation. It is important to show the patient the chances for a change in the situation. The patient must be guided to accept the handicap but also stimulated to make the best of it, to use all opportunities available and to cope psychologically with it.

Measures for Care

In 1994 the law mandating a compulsory care insurance was approved by the German parliament. It provided that the insurance pay the costs of care for frail and disabled persons of all age groups. It is financed, like other German social security systems, by a deduction directly from the salaries of those who are in the workforce and form the pensions of the retirees. Family members who are not in the workforce are included in the insurance of the spouses or parents. Like health insurance, unemployment insurance and pension funds, employers and employees pay half of salaries and pensions. Since July 1, 1996, it has been 1.7 percent. The insurance has paid for care in private households since April 1995 starting July 1996, the insurance has also paid for nursing home care.

The political goals of this proposal:

- to strengthen the care by family members in private households, and
- to lower the dependency of nursing home residents from social aid payments, thus saving the money of local authorities.

Although most political groups in Germany support these goals, the value of the first one point must be discussed from a gerontological point of view. Care in private households leads to many psychological and physical problems especially when persons with a higher degree of dependency are involved. Often the ecological and technical conditions are not sufficient and the family members lack a basic knowledge of care. Fifty percent of caregivers are women 65 years and older.

Twenty-five percent of care giving family members are women, 75 years of age and older. Clearly, family care has its limitations.

The most problematic points of the proposed care insurance in my view are:

- It seems likely—and the first care insurance statistics seem to prove it—most persons choose the financial support (70-80 percent) and not the service of district nurses, of professional care givers in the private households. The result is a financial improvement for families but not necessarily a better situation for the frail elderly.
- The incentive for rehabilitation measures is relatively low because the higher the level of care, the more money to which a family is entitled. Successful rehabilitation means less insurance money.

Care for the frail and dependent elderly can be regarded as one of the great challenges of our time. But everything should be done to preserve the greatest possible quality of life for dependent persons. On the other hand, we should try to avoid dependency, by extending the opportunities for intervention, prevention, and rehabilitation. These are tasks for interdisciplinary cooperation, to which biology of aging, geriatrics, behavioral as well as social sciences should contribute. However, this also is a challenge for policymakers and practitioners.

SUPPORT BY CARE INSURANCE

Persons Living in Private Households

- Level I: Persons who need care at least once a day (minimum 90 minutes) 400 DM resp. 250 US $ or services by nurses up to 750 DM
- Level II: Persons who need care at least three times a day (min. 3 hrs) 800 DM resp. 500 US $ or services by nurses up to 1800 DM
- Level III: Persons who need permanent care, at day and night (min. 5 hrs) 1300 DM resp. 750 US $ or services by nurses up to 2800 DM

Nursing Home Residents

- Up to 2800 DM resp. 1250 US $ contribution to the necessary payments for care.
- The costs for food and lodging are borne by the nursing home resident privately.

WE NEED DIFFERENT KINDS OF SERVICES FOR THE ELDERLY

Services to Promote Active Aging and Potential of Age (prevention)

- Institutions which support initiatives of the old for the young to give the elderly tasks and aims (helping by challenging, training).
- Cultural facilities and companies offering educational programs to prepare for retirement and aging.

- To offer stimulating opportunities for a large range of interests (cultural events, visits to museums, theaters and travel).
- Cultural facilities and universities which give the aged the opportunity to attend lectures and seminars.
- Sports clubs which offer a variety of sports for the elderly.
- Medical institutions for early diagnosis and treatment of possible risk factors and diseases.

Services for Dependent Aged Persons and Maintaining Independence in Aged Who Live Alone

- Creating supportive environments (to introduce necessary changes at home (installation of railings, roller-shutters, etc.)
- Using technical aids for maintaining independence (prosthetic function) and for increasing safety
- Senior citizen's clubs
- Day Centers for social communication and sports activities for the dependent elderly
- Setting up telephone-networks for social communication and for emergency calls
- Institutional support in:
    - personal care
    - preparing meals/meals-on-wheels
    - house cleaning/laundry services
    - shopping
- transport (for example, to day centers, etc.)
- mobile library
- escort and wheelchair services
- recreation/ recovery measures
- ambulant rehabilitation programs by a team of specialists

Centrally Located Community-Based Services for the Aged

- Cultural facilities for education
- Senior citizen's clubs
- Centers for initiatives and activities of the elderly and for social communication
- Day centers, day care centers
- Hospital-affiliated facilities (for geriatrics and psychogeriatrics), therapy centers
- Night clinics
- Rehabilitation in (geriatric) hospitals; interdisciplinary rehabilitation programs
- Homes for the aged offering part-time care
- Holiday accommodation in homes for the aged
- Self-help groups
- Institutions for psychological and social counseling

Services for Dependent People Cared For By Their Families

- Improvement of housing conditions
- Rehabilitation at home, cooperation between medical doctors, nurses and family members
- Rehabilitation in hospital-affiliated facilities
- Physical and cognitive training as a part of intervention programs
- Social services for caregivers (support in caring, preparing the meal, cleaning the house)
- Services providing substitutes for holiday-making caregivers daily/hourly visitor services supporting caregivers ("elderly sitting")
- Psychological support of the caregivers
- Counseling and self-help groups for caregivers
- Ambulatory nursing services (bathing aid, etc.)
- Part-time care in homes for the aged
- Medical treatment (including psychological support) of the caregivers for prevention of physical and psychological illness
- Financial support of the caregivers
- Additional help for the dying family member

Services for Dependent People In Need of Care, Not Supported by Family; People in Homes for the Aged

- Creating new concepts of homes for the aged. Different forms of temporary institutionalization/hospitalization; house-sharing accommodations; sheltered housing (with partial help and care available)
- Homes for the aged; geriatric departments of hospitals; nursing homes
- Counseling of the dependent on the choice of optimal institution in his/her given situation and his/her given care
- Anticipation of the new housing conditions
- Integration of rehabilitation programs in the care of the elderly. New concepts of physical and cognitive training
- Family programs for residents/inpatients
- Motivation of relatives for visiting the home for the aged and for sharing common activities with the family member (focussing the responsibility of relatives)

CONCLUSION

The longevity revolution is a challenge for us all. A policy for the aged, however, should not be determined only by the question: "What can we do for the aged?" We should also ask: "What can the aged do for society?" For this we need to revise our negative image of the aged. Most of the elderly are competent and wish to be engaged in activities on behalf of other people, for the good of society. Society, communities, and clubs are examples of the venues that can be used to promote the concept of voluntary public engagement. They can provide the framework to actualize the potential and services of older people.

If one listens to certain discussions about the economy one might get the impression that longevity is the worst thing that could happen to the human race because of the rising costs of pensions and of the health care system. Listening to some of the messages communicated one might get the feeling that it is unfortunate that so many people survive into great old age and get their pensions for which the younger generation has to pay. (*Conclusion:* It seems to be the duty of those 75 and older to walk into oncoming traffic.)

This certainly would be a wrong conclusion to draw from research on aging. One should not forget that it has taken humanity much effort over many centuries to enable us to survive into old age. The history of medicine is one of continuous efforts to prolong human life. It would be strange to view the great progress made in clinical medicine, or biology or pharmacology as bad luck for women and men. For many years WHO has promoted programs for healthy aging. Science and politics are striving together to create new medical and economic conditions to meet this goal—and to educate people in the importance of a healthier lifestyle.

The increasing life expectancy shows that they were successful. We should be happy. It would be ungrateful not to appreciate the progress made so far in attaining a remarkable decrease in morbidity and mortality and to focus only on the high costs associated with longevity.

We should regard the longevity revolution as an important gain and be ready to discuss the positive as well as the negative aspects in an unbiased fashion. Longevity should not be seen as a problem but as an opportunity and a challenge—a challenge for everyone: for the aging individual, for the family and for our society. We should not only focus upon the problems and deficits of aging and old age. We have to open our eyes to the new potential of the aged. We have to initiate research in this area, and also look to maximize the competence and potential of the very old. We have to see the aged as human capital. Eventually I am sure we will find new potential in the frail and disabled elderly as well.

MAINTAINING PROSPERITY IN AN AGING SOCIETY

Nick M. Vanston

Organization for Economic Cooperation
and Development (OECD)
2 rue Andre Pascal
Cedex 16
Paris 75016
France

INTRODUCTION

The implications of aging for OECD societies are deep and pervasive. The effects have been positive in recent decades. However aging trends are likely to result in a smaller proportion of the population being employed in the years after 2010, posing major challenges. There is likely to be reduced growth in material standards of living. Fewer workers to support more retirees raises fiscal issues and issues of inter-generational fairness. Increasing periods of life spent out of contact with the labor market in retirement raises fundamental issues about the allocation of work and leisure over life.

DEMOGRAPHIC CHALLENGES AND POLICY RESPONSES

Responding to these challenges requires action on many fronts. Public pensions, health and long-term care costs must each be contained. The structure of retirement income must be reformed and incentives to early retirement eliminated. People must be supported as they grow older, to enable them to continue to play a productive role in the labor market and in society. Financial market reforms are required in response to the huge growth in pension funds. Reforms that result in higher economic growth would alleviate the pressures of aging.

Strategic frameworks need to be in place at the national level in order to harmonize and sustain aging reforms and to build public understanding and support. A set of seven policy principles have been developed to guide these reforms, which are applicable with different degrees of urgency in different countries. International cooperation will facilitate the success of the reform process. Because many of the

Longevity and Quality of Life, Edited by Butler and Jasmin.
Kluwer Academic / Plenum Publishers, New York, 2000.

reforms require advance notice and gradual implementation, now is the time for action.

Aging Trends: A Smaller Proportion of the Population is Likely to Be Employed

In 1960, the older population aged 65 and over accounted for 15 percent of the working-age population. By 2030, this could jump to 35 percent, with particularly strong growth after 2010. In the past 25 years, the number of people of retirement age (65 and over) in most OECD countries (with the exception of Mexico and Turkey) rose by 45 million, but the working-age population rose by 120 million. As a result, population aging has so far posed no major economic or social problems for our societies. However, the effects of declining fertility and the aging of the baby boom generation will have significant implications for policymakers in the future. In the next 25 years, the number of persons of retirement age will rise by a further 70 million, while the working-age population will rise by only five million.

Not only are there more people in retirement age groups, but people are also retiring earlier and living much longer. For example, men in 1960 could expect to live some 68 years with 50 of those years spent in employment. Today men can expect to live 76 years, with only 38 years spent in employment.

The combined effects of all the aging phenomena (population aging due to fertility declines, individual aging resulting from increasing longevity, and changing labor force participation within age groups) is that the percentage of the population who are employed has been growing, and will continue to grow until about 2010, when it will start to fall. A reversal of trends toward early retirement would be the only major way to keep the employment ratio from falling.

A Smaller Percentage of the Population At Work Has Large Social and Economic Consequences, Including Reduced Growth in Material Living Standards

The goods and services produced by an economy largely determine the material living standards of a society. The quantity of goods and services produced depend mainly on the number of people working, and on their productivity. If productivity continues at its post-1973 average growth rate of 1.5 percent annually, the decline in the proportion employed would mean that the growth in material living standards will be cut in half in the decades after 2010.

A Different Allocation of Income Across Generations

When baby boomers start to retire, in about ten years time, the work force will decline. As time goes by and more people retire, we will see a smaller number of workers producing goods and services for an increasing number of retirees. If the balance of living standards between workers and retirees is to be maintained, more money will have to be put aside during working years, either through individual savings or pay-as-you-go taxes and contributions, in order to support longer periods of retirement. In 1960, for every older person there were over four employees to provide support. Currently, there are about three employees for each older person and that will continue for another decade. After 2010, the trend changes sharply. By 2030, the size of the older population will grow to 50 percent. That is, there will be only two employees for every older person.

People Living Longer and Healthier Lives But Retiring Earlier and Earlier

There is no economic or biological basis for retirement when people are in their 50s or 60s, yet the length of leisure in retirement is increasing rapidly. Policy must respond on several fronts.

Critical Policy Issues for the Future Arise as a Result of the Inevitable Aging of Our Societies

- will it continue to be possible to share societies' resources between the working generation and its dependent non-working members in ways that do not give rise to unacceptable societal and inter-generational conflicts?
- how can the contribution of older people to society and economic prosperity be enhanced?
- how should pension, health and long-term care best be reformed?
- what changes in the financial infrastructure are needed to support the development of advance-funded pension systems?
- to what extent will aging OECD countries be able to improve their well-being through growing trade in goods and services and assets, in particular with younger, faster-growing non-OECD countries?

Spending on Public Pensions, Health and Long-term Care Must Be Contained

In most OECD countries, sharing resources between working people and retirees involves public pension systems, which are usually financed by payroll taxes. Even though many Member countries have already taken steps to reform their public pension systems, these are insufficient to cope with anticipated future demands. The public pension accounts in most Member countries will start to go into sustained deficit in about ten years time. Public provision of health and long-term care for retirees will add to the burden.

Countries could finance future social spending obligations by raising payroll taxes to whatever level was necessary, but these would be so high as to discourage work effort and would cut deeply into working people's living standards. These considerations point to the overriding importance of curbing the growth of spending on public pensions, health and long-term care.

The Structure of Retirement Income Must Be Reformed

The provision of income in older age should take into account all the resources available to older people including public and private pensions, earnings and assets. Existing public pension systems which enable older people to maintain adequate standards of living are likely to remain the major source of old-age income for many retirees for some time to come. However, other sources may have to play a growing role, with the objective of spreading the burden across generations and enabling individuals to diversify risks across the different sources of retirement income. Such reforms are likely to mean that middle and high-income earners will want to supplement their public pensions. Hence, it will be important to establish a sound regulatory framework for private pension funds, including occupational pension schemes.

Reforms along these lines would change the current "implicit contract" between many governments and future retirees. Reforms must therefore be implemented with enough warning to allow people time to adjust to the new "contract," and to begin to anticipate problems likely to arise two or three decades down the road. In addition, there are limits on the speed that a country can move to an advance-funded scheme because of inter-generational equity considerations: current workers will be "paying twice," into their own pension fund and for the pensions of the currently retired.

Population Aging Will Put a Premium on More Effective Health and Long-term Care Spending

People are living longer and healthier lives. Nevertheless, population aging means that health and long-term care costs are likely to rise, although perhaps by less than was once feared. The central challenge is to ensure that these expenditures are cost-effective and meet the most pressing requirements, reducing time spent in dependence and in chronic care. This means that medical research and technology should be focused on the reduction of dependence arising from conditions which particularly afflict older people, such as senile dementia or arthritis.

Care-giving for frail older people is now often fragmented and unnecessarily costly. As demographic trends point to a particularly large growth in the numbers of people in the oldest age groups, it will be important to develop explicit policies and financial arrangements for care-giving that deliver quality and cost-effective services.

Incentives for Early Retirement Should Be Eliminated

Material standards of living, and hence the tax base, would also be higher if people worked longer. It is not a question of "forcing" older people to work longer. The current trend to early retirement is, in part, a reflection of a rising demand for leisure as societies become more prosperous, as well as a response to high and persistent unemployment. But current public pension systems, tax systems and social programs interact to provide a *strong* disincentive for workers to remain in the labor force after a certain age. Removing these disincentives, perhaps even providing positive incentives to work longer, coupled with effective steps to enhance the employability of older workers, could represent an important contribution to sustaining the growth of living standards.

But an increased willingness on the part of older workers to work longer must be matched by a sufficient number of job opportunities for them if higher unemployment is to be avoided. This, in turn will require a major change in the attitude of firms toward hiring and retraining older workers. Since these changes must be reflected in wage and labor cost structures, the cooperation of the social partners could play a very useful role in this process.

"Active Aging" Should Be Encouraged

"Active aging" is the capacity of people, as they grow older, to lead productive lives in society. One example of active aging is a flexible work-retirement transition. Active aging implies a high degree of flexibility in how individuals and families choose to spend their time over life—in work, in learning, in leisure and in care-giving. Public policy can foster active aging by removing existing constraints on life-course flexibility. It can also provide support that widens the range of options available to individuals via effective life-long learning or by medical interventions that help people maintain autonomy as they grow older. Indeed, the available evidence shows that the more active older people are, the better the quality of life they enjoy.

Significant Financial Market Reforms Are Also Necessary

Aging will change saving and investment patterns nationally, including the build-up and subsequent spending of private pension assets by the retiring baby boom generation. Differences in aging patterns across countries, particularly between the OECD and the non-OECD regions, will give rise to shifts in patterns of savings, investment and international capital flows. A more efficient global allocation of savings and risks could raise productivity, and therefore the goods and services available to consumers in OECD countries as the baby boom generation retires. The consequences of these global economic linkages for living standards in OECD countries are likely to be positive, but perhaps not very large.

- strengthening of financial market infrastructure through improved legislation and codes of conduct and the introduction of rigorous rules of transparency, fiduciary responsibility and disclosure;
- improved supervision, an upgrade and modernization of pension asset investment rules, and better domestic coordination among the different regulatory and supervisory agencies involved in the provision of retirement income;
- structural reforms in emerging market economies to ensure efficient allocation of savings into the most productive investment opportunities in the mutual interest of OECD and younger, faster growing non-OECD economies.

Higher Economic Growth Will Alleviate Pressures

Higher economic growth will alleviate pressures arising from population aging. Taxes to support retirees will be easier to bear when the working-age population can look forward to rising prosperity. Encouraging people to work longer will raise economic growth, increase the tax base, and reduce the numbers of dependent older persons. The higher national saving, if used productively, will result in higher capital stock and thus more output over the long term.

Although higher productivity is central to increasing living standards of the whole population, it does not directly address all of the economic, financial and social policy aspects of aging populations discussed above. Nevertheless, while an increasing proportion of national output will need to be transferred to retirees, the greater the economic growth, the easier it will be to ensure that making one group better off does not entail making another group worse off.

CONCLUSION

Populations of OECD countries are on average the oldest in the world. However, aging is taking place in all countries. Indeed, the relative importance of the world's older population has been growing for centuries. What is new is the rapid pace of change. In the non-OECD world, the rate of growth in the number of older people is much faster than in OECD countries. Nevertheless, the percentage of older people is still relatively small compared with the OECD, especially in Africa and southern Asia, because fertility rates began to decline only after child mortality rates fell. China, however, will soon become an older country.

The aging of populations, and policy reactions to it, will have direct and indirect implications for current balances and the international flow of goods, services and capital, both within the OECD area and between OECD and non-OECD countries. The consequences for living standards in OECD countries are likely to be positive, although perhaps not great.

THE ORGANIZATION AND FINANCING OF SOCIAL PROTECTION

Bob Williams

U. S. Department of Health and Human Services
Humphrey Building – Room 424E
200 Independence Avenue SW
Washington, DC 20201
USA

INTRODUCTION

It is a pleasure to be able to speak with you today. One of the major successes of the 20th century is increased longevity and improved health among older populations in both the industrialized nations and, increasingly, in many parts of the developing world. A related success is that persons who are born with a disability or later acquire one also have a better chance of leading a productive and satisfying life than ever before in history.

The aging of the population is the result of a long-term decline in fertility rates and an enormous improvement in life expectancy. In 1990 the number of persons over 65 in the U.S. was an estimated 32.0 million, or 12.3 percent of the population. By 2030 the elderly population is projected to grow to 22 percent of the population.

Other countries, of course, are experiencing even more significant increases in the sizes of their elderly populations. The very old, aged 85 and older, will be the fastest growing segment of the senior population into the next century. For example, in 1994, 3.7 million persons in the U.S. were aged 85 and older.

The U.S. Census Bureau projects that in the year 2050 the number of persons aged 85 and older could range from 9.6 million (low series) to 18.2 million (middle series) to 31.0 million (high series). Some leading demographers consider the Census' high estimate to be too conservative.

Consistent with the fact that older persons are living longer because of improved health, there is growing evidence that today's population aged 65 and older in the U.S. and other countries is less disabled than earlier cohorts. Compounded over long periods, relatively small declines in disability rates at older ages can have very significant implications for the number of seniors with

Longevity and Quality of Life, Edited by Butler and Jasmin.
Kluwer Academic / Plenum Publishers, New York, 2000.

disabilities and the costs of their and others' health care and long-term care services.

However, the overall trend in disability rates among younger groups is less cheering. Among persons at older working ages—those between the ages of 45 and 64—disability rates in the U.S. remained roughly constant from the mid 1970s through the early 1990s.

Among younger adults (aged 18 through 44) and children (under age 18), disability rates appear to have increased in the 1990s. Before 1990, disability rates for children and younger adults had remained relatively constant for nearly two decades. These figures indicate the need to focus on disability not only among the elderly, but to look at disability across the life span.

ACTIVE AGING

At the 1997 Denver Summit, the leaders of the eight largest industrial nations mandated a focus on "Active Aging." The Summit leaders opposed the portrayal of seniors solely in terms of the financial burden they represent for younger generations. The 1998 Birmingham Summit reiterated this theme.

It is noteworthy that in the U.S. the current emphasis on active aging is similar to the now truly global independent living movement first begun in the 1970s by younger persons with disabilities, both in America and throughout Europe.

Older people and younger people with disabilities share common interests. They refuse to be seen as "dependent." Rather, increasing numbers of individuals of all ages with disabilities are seeking choice and control over their everyday lives and their futures as well.

Changes in culture, technology and universally-designed environments are undermining what have long been held to be self-evident truths but in reality are false stereotypes. To make full use of the interests and capabilities of persons with disabilities, regardless of age, will require a fundamental rethinking of our approaches to social insurance, health care and other "safety net" programs.

STRATEGIES FOR ACTIVE AGING

The following strategies provide a foundation for fostering a climate of active aging in all our countries:

- Investments in age-related demographic, biomedical and behavioral research
- Evaluation of new health care interventions for persons with disabilities
- Assessment of health and retirement policies on labor force participation by younger and older persons with disabilities
- Support for, and recognition of, family care giving and volunteerism
- Modified living and work settings in support of persons with disabilities
- Enhanced consumer choice and control

Investments in Aging-related Research

Currently, the data needed to support collaborative research among the industrialized nations is underdeveloped. Many countries rely on aggregate data and models for making policy decisions. For some purposes, these are adequate; for others, they are not.

More longitudinal person-based survey data on the characteristics and behavior of older individuals and their families are needed to model and project population and disability trends and the use of health and long-term care services. For example, if surveys like the U.S. Health and Retirement Survey and National Long-Term Care Survey were replicated in other countries, they could provide valid cross-national comparisons on a host of age-related issues like:

- health and disability status
- use of services
- expenditures for health and long-term care
- economic status
- kinship patterns
- intergenerational transfers
- subgroup characteristics by gender and minority status

Evaluation of New Health Care Interventions

According to a 1996 WHO report, *The Global Burden of Disease*, non-communicable chronic diseases have replaced infectious diseases as the number one source of deaths in the world. Options for addressing chronic illness are growing, from high-tech treatments such as coronary artery bypass surgery or angioplasty, to low-tech solutions such as home care and assistive devices.

There should also be comparative research on entire health care systems to evaluate their overall performance—with emphasis on cross-national collaboration and the development of measures of disability that cover the full life span. We also need to increase public awareness of the importance of prevention, better nutrition and exercise in maintaining health, delaying the onset of disability and reducing its severity.

Labor Force Participation of Seniors

Although life expectancy has increased and the health status of seniors has improved over recent decades, in most developed nations labor force participation rates of men aged 65 or over have dropped significantly. Pension systems in many industrial countries, for example, "tax" work after a certain age, thereby creating disincentives to continued labor force participation.

For example, in the United States, Social Security beneficiaries with earnings above an "exempt amount" lose $1 in benefits for each $3 earned, thereby creating a disincentive to continued or renewed labor force participation. These shifts in labor force participation rates will reinforce rising senior support ratios as measured by the ratio of "retirement age" persons to the number of "working age" persons in coming decades. The industrialized nations could identify the most important disincentives in pension policies and health care coverage to continued work, as well as examine the effectiveness of incentives for prolonging productive activity.

Support for Family Caregiving and Volunteerism

Family caregiving is the key underpinning of the U.S. long-term care system. Healthy elders taking care of their disabled spouses and "young-old" children in their 60s looking after disabled parents 85 and older are able to keep many individuals at home who would otherwise require expensive institutional care.

Additionally, a quarter of Americans over 55 years of age participate in formal volunteer work of one sort or another—most often in connection with a church, a medical facility (e.g., hospital or nursing home), or service organization. Unfortunately, we do not always recognize the value of what volunteers contribute to society. A woman or man who cares for a spouse with a disability is not considered "productive." If that person were paid, the activity would suddenly become part of the Gross Domestic Product (GDP).

Modified Living and Work Settings

It is clear that we must foster the independence of seniors and younger persons with disabilities by providing appropriate supports to living and work environments. This requires greater focus on redesigned work settings, retrofitting of existing homes, new models of residential care, altered transportation systems, and a variety of assistive technologies. Together, these can help people overcome functional limitations and foster greater independence among all persons with disabilities, regardless of age.

Technology has an integral role to play in enhancing the independence of both seniors and younger persons with disabilities. For example, in the U.S. there is evidence from the National Long-Term Care Survey that older persons with functional limitations over time are relying less on informal care and more on assistive technologies.

Our policies and programs should foster, and certainly not impede, the development of new assistive technologies, increased access, and training and support for persons with disabilities. We tend, of course, to be more concerned with how assistive devices, ranging from canes and wheelchairs to very high tech devices, can assist persons with physical disabilities. One challenge is to develop better assistive technology and adapted environments for persons with cognitive disabilities, such as those with Alzheimer's or related disorders.

Enhanced Consumer Choice and Control

For both younger and older persons with disabilities, a new paradigm is emerging, based on the philosophy of personal choice. "Personal Assistance Service" (PAS) refers to assistance that enables persons with disabilities to accomplish tasks that the individual would normally do for himself or herself but for the disability. In the U.S., the notion of PAS grew out of the Independent Living Movement for persons with disabilities. The chief aim of the Independent Living Movement, started essentially by baby boomers in the 1960s and 1970s, was to ensure that persons with even the most severe disabilities exert maximum personal choice and control over their daily lives.

The connection between the Independent Living Movement and the emphasis on Active Aging is clear. Older persons are demanding to be viewed as competent, active and capable of contributing to their own well-being and that of others.

The PAS concept goes beyond traditional long-term care in several ways. First, it includes people of all ages, not just seniors. Second, it is not restricted to human assistance, but includes animal assistance, assistive devices, and environmental adaptations (e.g., curb cuts, specialized transportation accommodations). Third, PAS places high priority on the ability of persons with disabilities to decide for themselves the amount, type and duration of assistance they require. In this respect it supports consumer empowerment.

President Clinton has endorsed the concept of consumer-directed Personal Assistance Services. Last September he met with leaders of the disability community. He indicated interest in exploring how the Medicaid program could be restructured to allow persons living in institutions to return home and receive personal assistance in their communities.

To indicate the bipartisan nature of this goal, House Speaker Newt Gingrich was the one who introduced the Community Attendant Services Act in the House of Representatives last summer, as a means of focusing the attention of federal policymakers on consumer-directed services.

We in the Department of Health and Human Services are currently engaged in precisely that task, in partnership with state-sponsored programs, disability advocacy organizations and other groups. Our challenge is to find a balance between issues of personal choice and control, and those of public accountability. In so doing, we recognize that people with disabilities are a diverse lot and that the appropriate types of consumer-directed personal assistance will vary widely depending on the age, disabilities and cultural background of the consumer.

Shifting the focus of programs like Medicare and Medicaid from professional-directed to consumer-directed services requires a change in our prevailing mind set. Instead of providing services for people, we would in a real sense be investing in them and placing more emphasis on their abilities than on their disabilities.

And that leads me to some concluding observations on the nature of our social protection programs. There will always be a need to provide the traditional "safety net"of social security and health and long-term care coverage for certain individuals. At the same time, this can no longer be the dominant approach nations take with respect to their younger and older citizens with disabilities.

Many of our social insurance and safety net programs were designed in the late nineteenth and early twentieth centuries. They may no longer be adequate for the demands of the twenty-first century. Instead of simply "insuring against" the processes and perceived consequences of disability, we need to consider a shift toward social capital strategies that foster Independent Living and Active Aging.

The goal is not just to provide care and social protection but to tap into the productive capacities of all persons, regardless of age, disability or other circumstances. Such an undertaking requires that all our countries learn from one another through collaborative research, sharing of information and careful and continual reexamination of our policies and programs.

Looking Through the Mirror of Old Age

LITERATURE AND MEDICINE: García Márquez' LOVE IN THE TIME OF CHOLERA

Anne Hudson Jones

The University of Texas Medical Branch
Institute for the Medical Humanities
301 University Blvd.
Galveston, TX 77555
USA

"Tho' much is taken, much abides; and tho'
We are not now that strength which in old days
Moved earth and heaven, that which we are, we are, "(1)

INTRODUCTION

Arguably the greatest novel ever written about aging, Gabriel García-Márquez' *Love in the Time of Cholera* (2) may be a challenging text for those who need to read it most: the young, the would-be rational, and the impatient. To say that many health care professionals fall into these categories is not to fault them but merely to describe them. Who being young can know what it is like to be old? Who trained in western scientific medicine dares not try to be rational? And who caught up in the task-oriented imperative of contemporary medicine can truly claim the virtue of patience? Even before managed-care initiatives so greatly increased the pressure, physicians were famously time-driven, trained to seek efficiency in all things, care of patients prominently among them. To such persons, the thought of reading a novel may seem a profligate waste of time. Why spend hours reading about what never happened? This question has been eloquently answered over the years by those who use literature in medical education. (3-9) For *Love in the Time of Cholera*, the best answer is that what this novel offers cannot easily be found in traditional medical texts about aging.

GERONTOPHOBIA

The beginning of *Love in the Time of Cholera* is brilliant, a thematic tour de force without which the novel's ending, though still beautiful, would lose much of its contrasting power. Although the setting is exotic—a Spanish colonial city in a South American country (like García Márquez' native Colombia) on the Caribbean Sea in the late 19th and early 20th centuries, (10,11) the main character in this first part of the novel is a doctor, a man of science and progress, who offers medical readers a familiar perspective as they enter this fictional world. Dr. Juvenal Urbino de la Calle represents the best of the European scientific tradition. Born in the New World, he studies medicine and surgery in Paris before returning to his native city to try to elevate its standards of public health and culture. Devoted to science and reason, he remains a devout Catholic.

The novel opens at dawn on Pentecost morning, as Urbino answers an urgent call from the police inspector to come to the house of Jeremiah de Saint-Amour, the disabled Caribbean refugee (and fugitive) who is Urbino's close friend and favorite opponent at chess. Stepping into the darkened house, Urbino immediately recognizes the scent of bitter almonds "the fumes of gold cyanide" that for him are inevitably associated with "the fate of unrequited love" (p 3). Gold cyanide has indeed been the means that de Saint-Amour has used to kill himself, although his love for his beautiful Haitian mistress is passionately returned. Not unrequited love but fear of old age has driven de Saint-Amour to this act of self-destruction. Many years before, he had sworn that he would never be old, that he would kill himself at the age of 60. When de Saint-Amour turned 60, in January of the novel's current year, he pledged to carry out his suicide the night before Pentecost, which is "the most important holiday in a city consecrated to the cult of the Holy Spirit" (p 15). As the date approached, he became increasingly despondent because he did not want to die. Even less, however, did he want to grow old. And as a rationalist and an atheist, he felt bound to keep his vow.

Deeply saddened by his friend's death, Urbino exclaims: "Damn fool . . . The worst was over" (p 4). At 81, Urbino is, presumably, old enough to know. From his remark, one might assume that he is a model of vigorous aging, untroubled by the mental and physical frailties that de Saint-Amour must have feared. This is far from the case, however. García Márquez does not romanticizes the inevitable physical deterioration of aging, and Urbino's medical experience and superb clinical eye make him an excellent recorder of his own physical and mental decline, which is far advanced (panel 1). Perhaps no single detail more poignantly captures Urbino's loss of physical force than the contrast between the potent "stallion's stream" of his urination as a young man and the "oblique and scattered ...fantastic fountain" (p 30) that he can no longer control or direct. Finally, he has accepted the humiliation of urinating sitting down, like a woman. For the past 5 years, Urbino's wife has even had to bathe and dress him as if he were a baby.

Although de Saint-Amour may not have known the extent of his friend's physical failings, he must have known of his waning mental powers, because Urbino could no longer play chess without writing down each move so that he could remember his strategy. Observing this decline in his friend might well have strengthened de Saint-Amour's resolve to avoid such a future, however much he was enjoying his life at 60. Famed for playing multiple games of chess at the same time without ever having been beaten, de Saint-Amour must have feared such mental deterioration for himself. Ironically, what Urbino is referring to when he says "The worst was over," is the end of sexual desire, which he had welcomed about two decades earlier—that is, when he was about de Saint-Amour's age of 60. There is, however, no reason to believe that de Saint-Amour would have shared Urbino's relief at freedom from sexual desire and capacity.

By the end of this eventful Pentecost Sunday, Urbino has mourned de Saint-Amour and coined the word "gerontophobia" to explain the cause of his death; celebrated the silver anniversary of an eminent disciple in the profession of medicine; seen catastrophic thunderstorms; heard rumors of destruction by fire; and fallen to his own death while trying to recapture his escaped Paramaribo parrot, perhaps an exotic incarnation of the Holy Spirit. Thus, in the same day, two of the city's most renowned rationalists have died: one from fear of aging; the other still in "terror of not finding God in the darkness of death" (p 41). Jeremiah de Saint-Amour's strange name reflects his character: his love was not strong enough to offset his fear of the ravages of old age. He has lost the last game of his life because, his mistress reports, he played it "without love" (p. 14). Urbino, who "had always opposed prescribing palliatives for old age" (p 8), has for years taken secret medicines every hour in his attempt to deal with the problems of his own aging body. As a result, he always carries with him "a little pad of camphor that he inhaled deeply when no one was watching to calm his fear of so many medicines mixed together" (p 8). And his sudden death prevents his receiving his religion's final rites.

PATIENCE

As this first section comes to a close with Urbino's funeral, another major character of the novel emerges, Florentino Ariza, who has been waiting for this moment for 51 years, 9 months, and 4 days. For more than half a century, he has steadfastly sustained his love for Fermina Daza, now Urbino's widow. After the funeral, unable to wait any longer, Florentino tells Fermina again of his "vow of eternal fidelity and everlasting love" (p 50) for her. Furious that he would so profane the memory of her husband, Fermina orders Florentino out of her house forever. In her grief, she prays that night for her own death, so that she may rejoin her husband, but when she sleeps, her dreams are of Florentino Ariza, whose love she so long ago rejected.

The four middle parts of the novel provide a long flashback that tells of the youthful love of Fermina Daza and Florentino Ariza, as well as the long years of Fermina and Juvenal Urbino's marriage. It is the early love of Fermina and Florentino, their forced separation by her father, and the ingenuity with which they managed to stay in communication with each other by telegraph that parallel the love story of García Márquez' parents. (12) To read this long stretch of the novel, some readers may need to develop a kind of patience that does not come naturally to them. Nothing that is required of readers, however, can begin to approximate the extraordinary demands on Florentino, who sustains his love for Fermina throughout the half century of her marriage to Urbino. His devotion is touching: once during those many years he happens to see Fermina's reflection in a restaurant mirror, and he cannot rest until he has persuaded the proprietor to sell him the mirror so that he can take it home with him.

Never robust or handsome, Florentino guards his health well as the years progress. His greatest fear is that he will die or become too physically frail to woo

Fermina when she is free again. By the time of Urbino's funeral, Florentino Ariza is 76; Fermina Daza, 71. Because even as a young man Florentino has always looked old, he is less troubled by the physical effects of his aging than are others, although he fought hard against his premature balding before he finally accepted, at 48, its inevitability and shaved his head. Shortly afterwards, he lost all his teeth and acquired dentures. Despite his physical ugliness and his seeming passivity, there is something about Florentino that draws women to him, and he has had hundreds of sexual partners over the years-some 622 long-term liaisons, to be exact, as recorded in his 25 notebooks titled Women. Nonetheless, he considers himself to have remained loyal —if not sexually faithful—to Fermina. What sets Florentino apart from other men, in addition to his ugliness, is a romantic belief in love and a willingness to devote his life to that dream. He uncritically reads the romance novels and sentimental literature of his day, and he masters the art of writing love letters. It was through his letters that he first wooed and won Fermina when she was a schoolgirl. Like Cyrano de Bergerac, he has over the years written letters for many others who find themselves inarticulate in the face of love. Once he even found himself in correspondence with himself, writing letters for both partners in a love affair. What rationalists such as de Saint-Amour and Urbino might scorn as self-indulgent sentimentalism somehow gives Florentino the faith to endure his years of waiting for Fermina without giving up hope. There is about him a quality of divine madness, and he is often described in the novel as seeming to be inspired by the Holy Spirit.

LOVE IN THE TIME OF CHOLERA

Florentino's patience and García Márquez' alternative vision of aging culminate in the novel's splendid ending. After nearly 2 years of persistent, patient wooing, Florentino persuades Fermina to take a voyage with him up the Magdalena River on a vessel named New Fidelity. In these final pages, García Márquez writes of sexuality in old age as few writers have—in unflinchingly realistic detail but without the disgust expressed by Fermina's daughter and son at the prospect of sexual love between septuagenarians. On the first evening of the voyage, when Fermina and Florentino finally touch each other's hands, they both realize at first that "the hands made of old bones were not the hands they had imagined before touching. In the next moment, however, they were" (p 329). But that same evening Fermina tells Florentino to leave, because she smells "like an old woman" (p 329). Alone, Fermina thinks of him not as he was in his youth, but as he is now, "old and lame, but real" (p 330), and she asks of God "that Florentino Ariza would know how to begin again the next day" (p 330). He does indeed.

When they first undress to make love, Fermina tells Florentino not to look at her body because he won't like it. When he looks, he sees that "her shoulders were wrinkled, her breasts sagged, her ribs were covered by a flabby skin as pale and cold as a frog's," but he sees her "just as he had imagined her" (p 339). His impotence on this occasion, Florentino blames on loving her too much. Once more he must begin again. And the next morning he returns to Fermina's room, ready to do so. This time their effort is successful, and nearly 60 years after their first youthful infatuation, Florentino and Fermina consummate their relationship. It is the first time in 20 years that Fermina has made love. But García Márquez remains the realist: their union is not the idealized one of romance novels. Florentino has been rushed and clumsy, and they are both left feeling disappointed and sad. Once the urgency of this act has passed, they do not try to make love again until "the inspiration" comes to them without their forcing it. Until then, they are "satisfied with the simple joy of being together " (p 341).

The title, *Love in the Time of Cholera*, is associated with many aspects of the novel. Urbino first meets Fermina because he is called in to attend her when she falls ill during a cholera epidemic. Her father fears she has cholera, but Urbino correctly diagnoses a minor stomach ailment instead. Florentino suffers physically from symptoms that look like those of cholera, green vomit and diarrhoea, when he suffers most from love. He has attacks as a young man, when he is first wooing Fermina, and he has a memorable attack the first time Fermina agrees to receive him in her house a year or so after her husband's funeral. It is a decidedly unheroic affliction—as is the chronic constipation that requires Florentino to take regular enemas. And at the end of the novel, Florentino devises the stratagem of having the river boat New Fidelity return from La Dorado, the furthest port on the Great Magdalena River, under the cholera flag to preserve Fermina's privacy. Cargo and all other passengers are transferred to another ship. The ruse works well because the river is full of floating corpses that may be victims of yet another cholera epidemic. Only when the New Fidelity reaches its home port does it encounter a problem: the health officials forbid it to dock or its passengers to disembark. At this moment, when the Captain's ingenuity fails, Florentino's final, joyous inspiration offers solution (panel 2). Frost is already settling on Fermina's features, but Florentino will keep her under the protection of his love and the cholera flag—"forever."

This is a complex novel, and critics warn against reading it gullibly or interpreting it straightforwardly as just a love story. (11, 13, 14) García Márquez has called it his most political novel.12 It is not difficult to see some of his political concerns, the exploitation and destruction of the New World by its slavish adoption of Old World values. On this thematic level, the characters take on allegorical significance. Juvenal Urbino, whose father and son are both named Marco Aurelia Urbino, represents the failure of European rationalism and religion to master nature in the New World. Science, medicine, and progress seem to have much of value to offer, but ultimately they fail before the immense vital wildness of this land. Urbino's fear of animals, aging, and death—all natural forces—cannot be resolved by his rationalism or by his Old World Catholicism. Both Urbino's children were conceived during extended trips to the Old World. His son dies without an heir, and his daughter bears no male children. Thus, the Urbino line dies out in the New World. Florentino and Fermina, by contrast, have the natural vigor and vitality of the New World. Their names reflect their harmony with nature. Florentino, a bastard son, is never formally recognized by his father, nor does he bear the name of one of the popes, as do all the other men in his father's family. Fermina and Florentino repeatedly defy religious and cultural conventions that do not suit them. Fermina's angry shout, "To hell with the Archbishop!" (p 29), during a quarrel with her husband early in their marriage, echoes through the years. And Florentino is completely indifferent to rumors about his sexual preference, even though these rumors have deeply disturbed his Uncle Leo XII and nearly prevented him from naming Florentino his successor as President of the River Company of the Caribbean.

On literal, allegorical, and symbolic levels, the novel prepares for Florentino and Fermina to embrace their aging and explore life "that has no limits" (p 348). No man could be less traditionally heroic than Florentino, but he fulfills his uncle's conviction "that human beings are not born once and for all on the day their mothers give birth to them, but that life obliges them over and over again to give birth to themselves" (p 165). It is this quality of knowing how to begin again, in whatever circumstance he finds himself, that makes Florentino more heroic in the face of aging than are men such as Urbino and de Saint-Amour.

Despite the critics' warnings, for many readers—and especially for medical readers—the greatest value of this novel is its exploration of aging as a time of continuing discovery. When "medicine has no hope, or anything / More to give," patients may conclude that doctors "have no idea / What hope is, or how it comes". (15) García Márquez shows us what hope is, and how it comes.

Reprinted, with permission, from The Lancet, Vol. 350, October 18, 1997, pp. 1169-72.

REFERENCES

1. Tennyson A. Ulysses. In: Hill R.W. Jr, ed. Tennyson's poetry. New York: Norton, 1971: 54.

2. García Márquez G. Love in the Time of Cholera. Transl. Grossman E. New York: Alfred A. Knopf, 1988.

3. Trautmann J. The wonders of literature in medical education. In: Self DJ, ed. The role of the humanities in medical education. Norfolk, VA: Bio-Medical Ethics Program Eastern Virginia Medical School, 1978: 32-44.

4. Moore AR. The missing medical text: humane patient care. Carlton, Victoria: Melbourne University Press, 1978.

5. Rabuzzi KA, ed. Literature and medicine: toward a new discipline. Lit Med 1982; 1: 1-111.

6. Downie RS. Literature and medicine. J Med Ethics 1991; 17: 93-96, 98.

7. Charon R, Banks J.T., Connelly JE, et al. Literature and medicine: contributions to clinical practice. Ann Intern Med 1995; 122: 599-606.

8. Hunter KM, Charon R, Coulehan JL. The study of literature in medical education. Acad Med 1995; 70: 787-94.

9. McLellan MF, Jones AH. Why literature and medicine? Lancet 1996; 348: 109-11.

10. Bell-Villada G.H., García Márquez: the man and his work. Chapel Hill: University of North Carolina Press, 1990: 192.

11. Fiddian R. A prospective post-script: apropos of Love in the time of cholera. In: McGuirk B, Cardwell R, eds. Gabriel García Márquez: new readings. Cambridge: Cambridge University Press, 1987: 192.

12. Conti MS. A new epic from García Márquez. World Press Rev October 1996: 60.

13. Buckhead, and mercy. Garden City, NY: Doubleday, 1969

14. Booker MK. The dangers of gullible reading: narrative as seduction in García Márquez' Love in the time of cholera. Studies in Twentieth Century Literature 1993; 17: 181-95.

15. Columbus CK. Faint echoes and faded reflections: Love and justice in the time of cholera. Twentieth Century Literature 1992; 38: 89-100.

16. Dickey J. The cancer match. In: Dickey J. The eye-beaters, blood, victory, madness,: 31-32.

ETHICS AND LONGEVITY

Jean Bernard

82 rue d'Assas
Paris 75004
France

Last year, I reread *l'Ecole des Femmes*. Molière's hero, Arnolphe, is an old man. His advanced age is the subject of derision. His young wife is unfaithful to him. The old man is 42.

At the end of the eighteenth century, life expectancy was about 27 or 28. Fifty years later, Balzac wrote a beautiful novel to demonstrate that a woman of thirty can still love and be loved.

Four events are changing the course of human societies now and will do so even more in the next century. The first event is extension of human life to 80 or 90 years. The second event is the extension of life expectancy without incapacity, with variations from one country to another, but perhaps such variations are also due to a different appreciation of these extensions. The third event is the confirmed difference between the sexes. Aged societies are in their vast majority female societies. The fourth event is the extreme diversity of outcome. Life's great inequity is neither fortune, riches nor poverty. Nor is it gifts or talents. It is the last 20 or 30 years of life.

On one side, there are dependent old people, either an encumbrance to their families who distress them, or inmates of old people's homes.

At the other end of the spectrum, we find lively and alert old people, teaching, working, painting masterpieces as did Titian and Claude Monet. Frequently they give others the benefit of their wisdom, as did Fontenelle, aged 99 years and 10 months. In response to a friend's congratulations and assurances that he would reach 100, he asked, "Why are you so pessimistic?" A little later as he lay dying, and was asked how he felt, he replied that he had "some difficulty in existing."

Disraeli, Prime Minister to Queen Victoria, perceived three phases in old age. In the first phase, the old man realizes that he is not as brilliant as he once was. In the second phase, his entourage and the old man are of the same opinion, and in the third phase, only the entourage notices any change.

The reactions of the elderly vary a good deal. Some are troubled by these changes, but others accept them.

Agrippa d'Aubigné wrote: "Let my winter's night sleep on. I love the winter which purges my heart of wine."

Paul Valéry, probably referring to his own mother, wrote:

Longevity and Quality of Life, Edited by Butler and Jasmin.
Kluwer Academic / Plenum Publishers, New York, 2000.

"I am very old, I live in an intermediate world, already almost poised for each moment or circumstance. My memory is a house almost complete; this magic house could vanish in an instant."

Valéry again:

"I have grown old, but as a crystal forms. Impurities are eliminated, the spirit is purified and made simple. As is seen a view from far away."

Ethical questions raised by longevity (which I have been asked to consider) depend to a great extent on the evolution of human societies. Very broadly, we can expect three major evolutions:

The first evolution is mostly borne in the minds of writers. For instance, in an admirable Hungarian work of fiction published around 1935, when men and women approach the age of 60, they gather together on a road and move away into the distance until they disappear from view. They are never seen again. This fiction is inspired by a refusal to grow old. Death is programmed and thereby the afflictions of old age are eluded.

The same feeling has been expressed by other writers. Maurice Blanchot, in *Thomas l'Obscur*, describes the death of a woman who wants to die while she is still fully alive. More than ever before, she needs youth and beauty. Anne wishes to be wholly alive when she dies, and to avoid the intermediate stages of distaste and rejection of life.

This cannot be observed outside the world of fiction. In fact, huge numbers of human beings do not accept death. They want immortality and they ask men of science to make that possible. That is the reason for the second evolution. This second evolution is dominated by hope in the progress of scientific research in various domains.

First, and probably foremost, is the field of genetics. The recent cloning episode is a good example of this appetite for immortality, as when the parents of a child suffering from a lethal disease wish to clone the child so they can have an identical offspring who will survive. There are attempts to prolong finite cell life, or to limit the decrease of new cells.

There is also the physiology of nutrition and the attempt to understand the connection between low-calorie diets and longevity.

Finally, there is the field of neurology. In spite of recent progress, our knowledge of the brain is still imperfect. In fact, death is the death of a brain. Different kinds of old age depend above all on the state of the brain. Discoveries that might result from research on the central nervous system would certainly have significant consequences. Research in these four fields has already been responsible for a longer life expectancy and the anticipation of life without impairment. Additional longevity may be on the horizon. But there will be a limit. As things stand, death and life are entwined. Immortality is still the prerogative of the Gods of Olympus who are eternally young.

Having dealt with the first two evolutions, we must consider the third evolution happening before our eyes, which I would like to view in its relationship to ethics.

The work of the National Consultative Ethics Committee for Health and Life Sciences, and other work done in the last 20 years on bioethics has helped to define and outline principles governing bioethics. I shall try and examine how these principles apply to questions raised by longevity, while pointing out, as did Kierkegaard, that the more you simplify ethics, the clearer it becomes.

First principle: respect for the person. There are two definitions for *person*. The first is genetic. Since the discoveries of Jean Dausset and of the 600 million combinations of the HLA system, we know that each human being is unique and different from any other. The second definition is neural; man is defined by his aptitude for learning, his aptitude for creation. The person is a biological singularity, a being defined by psychosocial relationships, a subject for jurists. But the person transcends these analytic definitions and can be regarded as a value.

This value is not always highly regarded where the elderly are concerned. It is true that when the aged are still strong and active, they command respect. A great physicist of our time plays tennis at the age of 96. In 1976, Professor Robert Debré, aged 94, spoke in honor of Claude Bernard at the Académie des Sciences for an hour without referring to notes.

But the fate of dependent old people is very different. In their own families, they are sometimes mistreated or treated roughly or uncaringly. Mistreatment may be active or passive. In nursing homes and old people's homes care is sometimes attentive and compassionate (and I know of some outstanding examples), but in other cases it can be indifferent and demeaning. At best, pious lies are their lot.

An old person who has diminished faculties is still entitled to respect as a person even if his ability is partly, or sometimes greatly, reduced. He must be respected. Respect for what is human in each human being is an unqualified obligation. It is not a matter of monopolizing responsibility, but of federating vigilance in accepting the diverse forms of age.

Thus, bioethics gives first priority to respect for liberty and respect for dignity.

If one only considers liberty, there is a risk with the elderly of setting too narrow limits to the implementation of the principle. Then dignity comes to the fore. Once seen, dignity does not impose any obligation. It is a value that we decide to recognize. Kant wrote, "That which is above any price, which therefore has no equivalent, has dignity." The dignity of the elderly, whatever their circumstances, must be respected.

Second and third principles: respect of knowledge, the responsibility of the scientist, incorporating two major areas: the elderly themselves, and research.

Studies undertaken in the last few years have clearly established that there is a relationship between the degree of culture and education a person attains and that person's longevity. Both life expectancy and life without incapacity, are stronger and more likely for the educated than for the ignorant. Resisting the aging process is connected to a person's level of education. The frequency of senile dementia varies from 1 to 10 for those who are highly educated and those whose schooling was brief. Although this is sad from an ethical viewpoint, it is a fact. Progress made in recent years will be of greater benefit to those who are already privileged. There is, therefore, a moral obligation to better educate and impart more knowledge to the masses.

Paul Valéry wrote in the *Cahiers*: "The snake eats its own tail, but it is only after prolonged mastication that it recognizes in what it is eating the taste of the snake. It then stops eating, but after a while as it has nothing else to devour, it continues. It can end up with its head in its jaws. It calls that a theory of knowledge." Paul Valéry does not reveal the age of the snake, but particularly if it is old, knowledge for the snake depends very much on external research.

Some very weighty ethical matters are raised by the relationship between medical and biological research and geriatrics. First, its importance must be stated. Appropriately, children's diseases have in the past been the subject of intensive research, which as we know, has been successful. An ethical paradox arises because biological and medical research is to a large extent responsible for the growing ranks of the elderly with which we are concerned. Many diseases have been defeated and above all there is a surge of medical prediction and prevention which will dominate the twenty-first century.

However, research on aging is sadly lacking or even absent altogether. The elderly have been neglected for much too long and are still being neglected. The lack of interest (not to say disregard) shown by major national and international entities for geriatric research is ethically wrong. Also unethical is the general inattention and indifference of authorities to this population revolution, this extension of life expectancy that is now occurring. An increase in the numbers of the old and the very old is one of the major events of our societies as we reach the end of the century. As one distinguished geriatrician wrote: *A new medical discipline is born.* Only research can inspire the methods and progress of this new medical science. It is an ethical duty to supply this research with the necessary resources. Mighty efforts are needed now. It is important that we gain a better understanding of the molecular mechanisms which govern life from the day we are conceived to the day we die. It is unethical to give the elderly harsh and groundless therapy. It is unethical to accept ignorance in geriatrics.

One must also emphasize the ethical importance of research on palliative care. There has been justified criticism of extending the use of palliative therapy methods to the young. For the elderly, palliative therapy is frequently justified and leads to an acceptable compromise which obviates two contrasting excesses—aggressive and futile therapy, and euthanasia.

It is also worth noting that progress in the nosology of senile dementia and the isolation of Alzheimer's disease have undoubtedly brought about improvements. The old and demented used to be treated with contempt and left to their own devices. To quote Moulias, someone suffering from Alzheimer's disease in our times is still a tragic figure struck down by severe illness, but a respected figure. It need hardly be said that when the very old spend their remaining days in a hospital, their association with the staff is of paramount importance. They need to be cared for lovingly and discreetly, at their own pace and with due regard to their wishes, which may vary from one moment to the next.

Other ethical issues need to be addressed when they involve the advanced age of the patient or his mental aptitude. One example is informed consent when the elderly patient is no longer able to give it. The family's attitude may not be guileless. Other possibilities must be sought, for instance, enlisting a magistrate. But perhaps more appropriately, the physician can shoulder this responsibility if he has been adequately trained to do so.

Fourth and last principle: the rejection of lucre or, more specifically, an examination of the relation between ethics and money. The first ethical battle was waged against iron. The commandment is ancient: “Thou shalt not kill.” And now ethical principles meet a new foe. This is not a golden age, it is an age in which gold rules the world. It is important to underline the dangers inherent in the present connection between profit and the human body.

In the North, in Europe, in the United States, in Japan, the situation has improved for the elderly. There remain inequalities between the old who are rich and the old who are poor, but they are limited.

As distinguished geriatricians have remarked, there is common ground for ethics, research and economics. The better the care lavished on the old, the less they cost society and the longer they stay independent and out of institutions.

Ethical deliberations can play a major role in this respect. They can mediate or try to satisfy both medical and economic imperatives.

However, (and this is one of the ethical scandals of our time) major inequalities persist between North and South. In the South, the elderly are frequently deserted by their own families. They are exposed to two great threats, malnutrition and infection. Of course, the situation varies with different countries. In Indonesia, there has been notable progress whereas stagnation reigns cruelly in Nigeria. This discrepancy between North and South is probably the most critical ethical issue at the end of this century. Very energetic efforts are required.

LONGEVITY AND QUALITY OF LIFE

Elie Wiesel

Boston University
745 Commonwealth Avenue
Boston, MA 02215
USA

My paper is in praise of longevity and is also a tribute to old age. When little Joachim-Alexandre, whom you've just seen, reaches my age, he will still be young.

But who decides when youth ends and longevity begins?

I can remember my grandfather. I was a little boy, he was old and I loved him. I loved him with all my heart. He was kind, gentle, generous, wise and handsome. He had the knack of being able to soothe me. I would talk to him in a way I never talked to anyone else, with absolute trust. I would tell him my secrets and I knew that he would never betray me, would never pass judgment. My grandfather was my protector and my ally, not my critic. And, to make him happy, I was prepared to do anything, to overcome the fiercest forces and the most hazardous obstacles. He was old and, in my eyes, why not immortal? It's because of him that, even as an adult, I feel such deep affection for old people.

In New York, for a very long time, I used to spend an afternoon a month, or sometimes every week, visiting old forgotten Yiddish writers. By listening to them, I made them feel that they still existed for the world, for the outside world, and, more importantly, that what they had written was important, and that their memory was still alive. Even today I love old people, but not old age. Everyone is trying to grow old without looking old. No one likes old age now; it's called longevity; that sounds better. Old age implies deterioration and decrepitude, the inevitable decline, moving closer to an irrevocable end; to use a biblical image, it is the return to dust. So we try to disguise death, adorning it with flattering terms: the "third age" for the French, or the "golden age" for the Americans. In other words, as you have heard today from our illustrious and generous confreres, by the next century, even at the age of a hundred, people will be dying young. But modern society has not shown great tenderness toward the elderly. Let's be frank and say it: society today is only interested in youth and has developed a cult of youth worship. Just look at advertising around the world. There's always a young man with dazzling teeth, embracing a young woman, her sensual body alive with desire. One might say that the young are today's gods. They see everything as their due and are given even more.

Longevity and Quality of Life, Edited by Butler and Jasmin.
Kluwer Academic / Plenum Publishers, New York, 2000.

There is a total paradox as medical science on the one hand is determined to extend the lives of men and women and on the other, they are stopped from enjoying it. Old people should simply stay at home and not bother anyone; they should be happy just being fed, clothed and kept warm. At best, they might be sent off to lie in the sun somewhere. In the States we send them to Florida, to rest, sending them into compulsory retirement. What a horrible term–retirement! In military terms it means retreat and defeat. Making them recluses, we make them feel useless, unnecessary, and victims of a system of perpetual humiliation; all they can do is feel shame for not being young and guilt for still being alive.

In the past, things were different. Shaped by memory, experience and learning, an elderly person inspired respect, obedience, admiration and gratitude. Old age was synonymous with wisdom. Even a young sage claimed to be old. In those days, it was an honor to be old. To quote an 18th century philosopher, "I love everything that is old: old-fashioned manners, old wine, old friends and old books." In earlier days, we would have added "old minds." In the Roman senate and the great academies of ancient Greece, it was the older figures who commanded respect and attention. In the Talmud, in both the Talmudic and Midrashic universe, the central figure was always an old man from whom all authority sprang. A society or civilization can be judged by the way it treats the elderly. According to the Scripture, respect must be shown to the elderly; according to the Bible, I must rise to my feet in the presence of an elderly person, must offer him my seat, listen to him and remain standing while I do so.

His knowledge is, *a priori*, greater than mine. His generation is closer to the source. An old person knows more than I because he remembers more things than I. To work for him therefore goes beyond duty; it is a special privilege. Helping those who cannot help themselves is one of the primary duties of a person with an ethical conscience; and you have just heard a brilliant speech from the man who, in effect, personifies ethics in France, Jean Bernard. But not all old people are equal. All those who need us are not equal. It is important to help a child, but that is also a paying proposition, a form of investment. One day the child will grow up and become rich and perhaps influential; he may express gratitude for our kindness on his behalf, and the faith we had in his future. But an old person . . . helping him to live, to avoid despair, offers no prospects of reward. To help such a person is therefore an act of grace, a totally altruistic and entirely, humanly pure act. Some of my colleagues here raised a question which I, in turn, shall ask: why do we now do so much for children and young people and so little for the elderly? I am not talking about our Council, which has done far more than anyone else. But why have we failed to do for some what we are trying to do for others? Our moral duty is to help children because they are young, and the elderly because they are not. In the child, I celebrate promise, but in the old person I celebrate memory. Surely we need one another if we are to reach fulfillment, to keep on living in a world which, in fact, until now, has had no real place for us.

When dealing with the elderly, why is so little care taken of both their physical and mental well-being? How can people forget so easily that one day, with a little luck, we shall be in the same situation? Does this perspective frighten us? To paraphrase Paul Valéry, who has already been quoted here today, we could no doubt say: "O, old man, now you know man is not immortal. We are afraid."

Among the characters in my novels, there is always a madman, a child and an old man. The madman is not really mad, the child is no longer a real child, and the old man is a kind of prophet, although he is a prophet not of the future but of the past. All three are penetrated by a beauty both serious and noble, and their smiles convey warmth and reassurance. What are they doing in my stories? I treat them as guests of honor there. They are bound by a special relationship which the enemy was able to detect it. They were the first marked out for death. Expelled from everywhere, repudiated by society, undesirable to the powers that be and doomed by the cruelty of the enemy and the indifference of friends, I can offer them my pages as a refuge. We saw children and old people walking together, hand in hand, in nocturnal processions, beneath a sky of incandescent ashes. After the Liberation a strange phenomenon occurred. While among other societies there were mostly children and their grandparents who had never been involved in the fighting, among the Jewish people, there were men and women in the prime of life. We had become a poor community, deprived of the older generation, dispossessed of the past; and the children of their future. In those times, children themselves were old people. So what does it mean to be old? When you are a child, you want to grow up: you become a teenager, then an adult and then grow old; and when you are old you return to childhood. Is a child powerless? Not necessarily. A child in the most poignant sense of the term. Like a child, an old person now, thanks to progress extending human life can also embark on the discovery of the world, rediscovering thought, which is the life-force of the world. And like a child, he is open to hope.

Of course the elderly today no longer live under physical threat to their existence. In the civilized world, their material needs are provided for, but they are alone, forsaken by their own families. What has organized society done to ease their feeling of loneliness? How can we permit ourselves to live so far from men and women whose need for our presence is so great and who have become the victims of a form of loneliness for which no one is directly responsible? Well, for those of us here today, things are quite clear. Confronted with human beings who have had the good fortune to enjoy the advantages and blessings of longevity, I believe that it is our responsibility to tell them, and in no uncertain terms, with a feeling of gratitude, that their happiness enriches ours, as their recollections will be part of it. If they are our past, aren't we their future?

In fact, on the scale of human eternity, without even contemplating divine eternity, the past is no shorter than the future. For all these men and women who are old–both future and present–let us be happy that they are, for without their light, ours would have been swallowed already by the darkness that comes not from God, nor from time, but from human beings and their inexhaustible and complacent indifference.

Conclusion

SYNTHESIS

Harvey V. Fineberg

Harvard School of Public Health
677 Huntington Avenue
Boston, MA 02115
USA

We have all benefited enormously over these three days from sharing with one another the perspectives that we brought as demographers, as scientists, as citizens, as academics, and as healthcare professionals concerned with the problem and the opportunity represented by the longevity revolution. I would be remiss if I did not begin this set of reflections with a tribute to the organizers and chairs of this symposium who made it possible for us to enjoy this time together and to learn so much; Claude Jasmin and Robert Butler. Thank you both very, very much.

In the part of the world where I live, farmers make maple syrup. They take about ten gallons of sap from the maple tree and boil it down to one quart of sweet syrup. I feel my task is something like that of the maple harvester. Instead of concentrating raw sap, I am trying to distill the essence from what is already a pool of very refined and very rich material. In these reflections, rather than attempt to summarize such a remarkable set of presentations, I would like to suggest three general conclusions that derive from our discussion and to say a few words about why I think each is so important.

The first conclusion is that we are in the midst of the most profound and unprecedented demographic transition in the history of humanity. It is reshaping the world's population, and it is affecting all of our lives. We heard the evidence about this transition from many quarters. What I think we perhaps did not emphasize sufficiently is that the demographic transition is a change both in the absolute number of older citizens and in the proportion of society that will be made up of older citizens. Both the growing number and the rising proportion of older persons have infiltrated our discussions, explaining apparently contradictory elements. On the one hand, the elderly will be more independent, autonomous, and powerful. At the same time, society will have to cope with greater numbers of individuals as a proportion of the workforce who are no longer working, who are a

Longevity and Quality of Life, Edited by Butler and Jasmin.
Kluwer Academic / Plenum Publishers, New York, 2000.

burden in terms of long-term and nursing care, and who in many ways require us to rethink the way in which we organize our institutions and our lives.

We talked about the origin of this unprecedented transition in its biological manifestations and in its social context. It was rather discouraging, was it not, to learn that in evolutionary terms, sex and death entered the scene at the same time? Before there was sexual reproduction, cells simply divided and survived indefinitely. With sexual reproduction, we were told, the genome alone is perpetuated and the soma (the rest of us) dispensable.

We talked about the origins of the longevity revolution in behavioral terms, in terms of the social conditions that have made it possible for people to live longer and healthier lives. We discussed the many ways in which the powerful combination of reduced mortality at younger ages that began early in the century is now being matched by an accelerating pace of survival amongst the older ages, giving rise to hope that the longevity of the species has yet to be fully realized.

A second major conclusion that I drew from our discussion is that it would be a mistake to think about this worldwide transition primarily in terms of numbers and proportions of the elderly. More significantly, it is a revolution in ideas and concepts, a revolution in the roles, capacities, and expectations at later stages of life. The hidden implication of this emphasis is that the revolution in longevity will bring change for all of us at every stage of life. We talked about the lessons to be gained from other social movements and philosophies - the women's movement, the new ways of thinking about the integration of persons with a variety of disabilities into mainstream society. We used terms like "active aging," "successful aging," and "healthy aging," to convey this fundamental idea of a change in the concept of what it means to be older today and tomorrow, as contrasted with the past.

I want to share with you a personal experience, which brought this home to me very powerfully. My wife, Mary Wilson, is a physician, an infectious disease specialist, who writes about the emergence of infectious diseases. A few years ago at Christmas, we were visiting her parents. They are among the one percent of Americans still living on family farms. They are situated in Indiana, about a hundred kilometers north of Indianapolis, outside of Kokomo, close to a small town called Greentown. Soon after we arrived, Mary's mother asked her how everything was going. "Fine," Mary said, "in fact, I've just published an article in the British Medical Journal about emerging infections. It's in a special December issue." "Wonderful," said her mother, "I'd love to see it." "I'm sorry," said my wife, "I couldn't bring it. It hasn't arrived at the library, and we didn't get our copy at home." "Well," her mother said, "I wonder if it's on the Internet?" And in a few minutes she was reading the article, downloaded via the Internet to the farm in Indiana, that we in Boston had not yet received. This story suggests to me not only the cleverness of my mother-in-law, but also the profound ways in which technology and change will contribute to the revolution in capacity and expectation of what it means to be an older citizen.

My third conclusion is this: the full realization of the opportunities resulting from the longevity revolution will depend on many actions that are, in principle, within our control. They are within our control as individuals and as family members. They are within our control as workers, professionals, scientists, leaders in civic, healthcare, and nonprofit institutions. They are within our control as citizens in democratic societies, as political leaders, as media figures, as intellectuals, as community advocates, and as individuals who will shape the society in which we all live. And the opportunities arise at every level. We heard about new insight into the genetic foundation of aging and senescence. We heard about the promise of research on telomerase and DHEA. We heard about the importance of prevention through health care and through public health measures, to prevent disease outright and to assure that disease, when present, does not progress. We talked about the needs and opportunities for institutional reform, not only within those institutions that deal directly with the elderly, but in institutions that relate to all of us, including educational institutions. In future years, the elderly as well as others may move from time of work, to time of education, to time of leisure in new ways.

The question, I think, for all of us is whether we will succeed in taking the necessary action. Will we have the scientific ingenuity? Will we have the personal discipline and sense of responsibility? Will we be sufficiently innovative with our institutions? Will we have the character and courage to act politically? Will we be able, in other words, to seize the opportunity that the longevity revolution represents and leverage that opportunity into a reality of a more successful society, not only in terms of successful and active aging, but successful as a social system? This is the deeper and lasting challenge of the worldwide revolution in longevity.

CONTRIBUTORS

ANHOURY, Pierre
Public Health Specialist
Medecine Sociale et Geriatrie
Associate Director, Health Division
Ernst & Young
Paris, France

APT, Nana
Professor of Sociology
University of Ghana
President, African Gerontological Society
Legon, Ghana

ATTIAS-DONFUT, Claudine
Director of Research on Aging
Caisse Nationale d'Assurance Viellesse
Paris, France

BAULIEU, Etienne-Emile
Professor, College de France
Director of Research, INSERM
France

BERNARD, Jean
President Emeritus
Comite National Consultatif d'Ethique
University Professor
Paris, France

BEZRUKOV, Vladislav
Director, Institute of Gerontology
Kiev, Ukraine

BIRREN, James
Associate Director
UCLA Center on Aging
Los Angeles, California – USA

BREEZE, Elizabeth
Chief, Conference on Epidemiology
London School of Hygiene and Tropical Medicine
London, England – UK

BUTLER, Robert N.
President and CEO
International Longevity Center-USA
Professor of Geriatrics
Mount Sinai Medical Center
Member, International Counsel for the Advancement of Global Health
New York, New York – USA

FINEBERG, Harvey V.
Provost, Harvard University
Vice President, International Counsel for the Advancement of Global Health
Boston, Massachusetts – USA

FOGEL, Robert William
1993 Nobel Prizewinner in Economics
Director, Center for Population Economics
University of Chicago
Chicago, Illinois – USA

FOLEY, Kathleen
Director, Project on Death in America
Chief, Pain Service
Memorial Sloan-Kettering Cancer Center
Professor, Neurology, Neuroscience and Clinical Pharmacology
Cornell University Medical College
New York, New York – USA

FORETTE, Francois
University Professor
Chief of Service, Hopital Broca
Director, Fondation Nationale de Gerontologie
President, l'International Longevity Center-France
Paris, France

FRIEDAN, Betty
Sociologist, Writer
Founder, National Organization for Women
New York, New York – USA

GUESRY, Pierre
Associate Director of Research
Centre de Recherche Nestle
Lausanne, Switzerland

HARLEY, Calvin B.
Scientist
Geron Corporation
Menlo Park, California – USA

HUDSON JONES, Anne
Professor, Literature and Medicine
Institute for the Medical Humanities
University of Texas
Galveston, Texas – USA

IACCARINO, Maurizio
Deputy Director-General
Sciences exactes et naturelles
UNESCO
Paris, France

JASMIN, Claude
Founding President
International Counsel for the Advancement of Global Health
Professor of Oncology
Chief of Service, Hematology
Hopital Paul Brousse
Director U 268, INSERM
Villejuif, France

JOYCE, Charles Richard Boddington
Professor of Psychology
Royal College of Surgeons
Dublin, Ireland

KIRKWOOD, Thomas
President
British Society for Research on Ageing
Professor, Biological Gerontology
University of Manchester
Manchester, England – UK

KOUCHNER, Bernard
Secretary of State, Health and Social Work
Minister of Employment and Solidarity
Paris, France

LAWTON, M. Powell
Director of Research
Philadelphia Geriatric Center
Philadelphia, Pennsylvania – USA

LEHR, Ursula
President
German Society of Gerontology and Geriatrics
Director of Science
German Center for Research on Aging
University of Heidelberg
Heidelberg, Germany

MANN, Jonathan M. †
Founder
Global Program on AIDS at OMS
Dean, School of Public Health
Allegheny University
Philadelphia, PA – USA

PILLER, Gordon
Founding Director
Leukemia Research Foundation
Treasurer, International Counsel for the Advancement of Global Health
London, England - UK

POMPIDOU, Alain
Deputy - Europe
President, STOA
Professor of Histology, Embryology and Cytogenetics
Universite Rene Descartes
Paris, France

REGNIER, Francois
Associate Director
Direction Generale Pharmacie
Synthelabo Groupe
Le Plessis-Robinson
France

SADIK, Nafis
Executive Director
United Nations Population Fund
New York, NY – USA

SCHOLER, Dietrich W.
Director
Pharma Policy and External Affairs
Novartis Pharma AG
Bale, Switzerland

STAHELIN, Hannes, B.
Chief
Clinique Geriatrique Universitaire
Bale, Switzerland

SVANBORG, Alvar
Professor Emeritus of Geriatric Medicine
University of Gothenburg
Gothenburg, Sweden
The University of Illinois
Chicago, Illinois – USA

TAYLOR, Humphrey J.F.
President
Louis Harris and Associates
New York, New York – USA

THAIRU, Kihumbu
Professor
University of Nairobi
Nairobi, Kenya
Member, International Counsel for the Advancement of Global Health
Kikuyu, Kenya

Van WEEL, Chris
Professor
Department of General Practice and Social Medicine
Universite de Nijmegen
The Netherlands

VANSTON, Nick
OCDE
Paris, France

VAUPEL, James W.
Director
Max Planck Institute for Demographic Research
Rostock, Germany

WIESEL, Elie
Writer
Nobel Peace Prizewinner
Professor of Religion and Philosophy
Boston University
Member, International Counsel for the Advancement of Global Health
New York, NY – USA

WILLIAMS, Robert
Assistant Secretary of State
Handicapped, Aging, and Long-Term Health Care Policy
U.S. Department of Health and Human Services
Washington, D.C. – USA

INDEX